# Natural and Herbal Beauty Care PENNY PINCHERS

GW00691654

# Contents

# Back to Nature

Nowadays beauty experts are concerned with good health as well as good looks. Clear skin, shining hair, a slim figure, healthy teeth and nails—they're all a vital part of the total beauty look. Cosmetics are formulated to enhance natural assets, rather than disguise them. More and more cosmetic manufacturers are including natural ingredients in their formulae—things like honey, herbs, fruit and vegetables.

The original recipes for many effective cosmetic preparations are based on old recipes handed down for generations. Your grandmother almost certainly knew more than you do about preparing lotions and potions from the most simple ingredients—and the likelihood is that her potions worked. It's now acknowledged that herbs, fruit, and flowers often contain healing, soothing or beautifying properties.

It's fun, and it certainly saves money, to experiment with some of these old recipes. Usually at least some of the ingredients are already at hand in the larder, such as olive oil, yoghurt, fruits and vegetables. Other ingredients such as rose-water, witch-hazel, fuller's earth, kaolin, etc, can be bought very cheaply. However, I don't consider that home-made cosmetics can totally replace commercial ones. There's no substitute for the latest lipstick, a good mascara or some of the super skin creams and cosmetic pencils now on the market. But, when money is tight, it's wasteful to spend precious cash on expensive items such as face packs, body oils, bath additives, and hair treatment creams when these are so easy to make and so effective.

If you have a sensitive skin, home-made products do have one distinct advantage over commercial ones : they are often much gentler. For instance, a facial made from avocado flesh is very soothing. If you are allergic to ingredients such as alcohol, aluminium salts, benzaldehyde, then using natural substances, like yoghurt, makes sense. But, remember it is possible, although rare, to be allergic to almost anything, including yoghurt, natural herbs, fruit and vegetables which can contain fairly concentrated chemicals such as the citric acid in lemons. So, do take care that you choose the correct recipes for your particular skin-type when selecting them from this book. And, if you develop an allergy, stop using the recipe responsible.

Since economy is the main idea behind this book, don't be tempted to rush out and buy out-of-season fruits and vegetables to use in the recipes given. The experienced pennypincher always uses the things that are to hand or cheap to buy. In most of the recipes, squashy fruits (the kind at the bottom of the basket which are usually thrown away) are just as good as unblemished produce. The herbs are easy to grow (see pp 62–63), and the binding ingredients, oil and eggs, are household basics.

Critics of natural cosmetics enthusiasts often say : 'Of course, all that edible "goo" you're plastering on your face would do you a lot more good if you actually ate it.' There is some truth in this ; diet is a vital part of any beauty routine, and correcting dietary habits can make a dramatic difference to skin and hair problems. The foods that do you good inside are usually the ones that do you good outside too. Every beauty diet mentions fresh fruits, vegetables, yoghurt, and whole-grain food such as bran and oatmeal and it's no accident that these same foods are used so much in the preparation of natural cosmetics. Some foods actually attack the same problem from two directions : lemons, for instance, are an excellent source of Vitamin C, vital for clear, healthy skin ; they also act as an astringent which can help clear your skin from the outside. Yoghurt, too, helps fight unfriendly bacteria both from without and from within.

Making your own cosmetics involves rethinking your beauty routine and rethinking the usefulness of various products and equipment. Model girls are the professional experimenters and they are ingenious at finding unusual uses for cosmetics. For instance, there's no reason why one lip-pencil shouldn't be used as cheek and eye colour too, softened where necessary with moisturizing lotion. I have included lots of money-saving ideas which will give your looks—and bank balance—a welcome boost.

# What You Need

I have deliberately avoided using recipes and ingredients which involve hours of seeking out and preparation. You won't find a double boiler, distilling equipment, gum arabic or strange-sounding herbs mentioned in this book. The most exotic utensil is a pestle and mortar (but the back of a spoon and a shallow dish will do just as well), and the most expensive ingredient is avocado (but bananas are almost as good). Most of the fruits, vegetables and other materials can be obtained from your supermarket or chemist. The herbs are probably available, dried, at your supermarket, too, although you may decide to grow them yourself. Here is a list of the ingredients mentioned and where to get them:

## Supermarket, Greengrocer's

### Fruit and Vegetables

Apricots, leeks, turnips, peaches, tomatoes, cucumbers, carrots, onions, grapes, lemons, avocados, apples, bananas, melons, potatoes.

### Larder Basics

Cloves, yoghurt (or you can make it yourself, see recipe on p 5), mineral water, corn, olive or almond oil, mayonnaise, cream, milk, yeast, honey, cider vinegar, eggs, porridge oats, cinnamon, carraway seeds, raspberry- and strawberry-flavoured jelly, ground almonds, peppermint, rum or almond essence, cochineal, teabags, oil of cloves, peppermint oil, bran, salt.

## Chemist's

Baby oil, lavender water, rose-water, witch-hazel, Friar's Balsam, glycerine, kaolin, petroleum jelly, lanolin, lavatory paper, sun oil, pumice, baby shampoo, fuller's earth, henna powder, oil of cloves.

## Supermarket, Health-food Shop, Herbalist

Parsley, mint, elderflower water, eyebright, camomile, rosemary, basil, sage, geranium or musk oil.

## Garden, Hedgerow

Nettles, marigold petals, nasturtium leaves, eyebright.

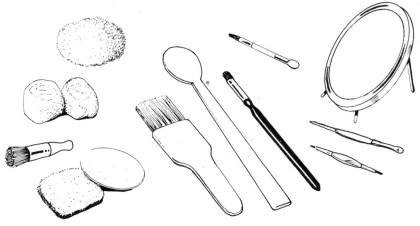

# Recommended Stockists

Culpeper Ltd is a leading herbalist and a treasure-house of fascinating herbs, oils, literature and gift ideas. Ingredients used in this book and stocked by Culpeper's are: elderflower water, rose-water, rose geranium oil, eyebright, parsley, mint, camomile (Culpeper's recommend the Belgian variety), rosemary, basil, and sage.

*Branches*

21 Bruton Street, London W1X 7DA
  (01-629 4559)
9 Flask Walk, Hampstead, NW3
  (01-794 7263)
14 Bridewell Alley, Norwich
  (Norwich 618911)
25 Lion Yard, Cambridge
  (Cambridge 67370)
The Friary, Grosvenor Centre,
  Northampton (Northampton 39288)
12D Meeting House Lane, Brighton
  (Brighton 27939)

All branches will send products by post.

Boots the chemists stock all the items listed under chemist's.

Jacksons:
171 Piccadilly, W1 (01-493 1033)
6a/6b Sloane Street, SW1
  (01-235 9233)

These two stores stock both dried herbs and a large range of essential oils (including geranium, lavender, elderflower, cinnamon and pine).

Safeway Food Stores are good for fruit, vegetables and many of the more unusual larder items such as peppermint oil, cider vinegar and oil of cloves.

## Yoghurt

Once you've made your own yoghurt, the sweetened shop-bought kind will never taste as good. Here's a recipe for making yoghurt—for use in the recipes in the book or just to eat. A yoghurt-making machine is fairly inexpensive to buy and a very worthwhile investment from both the cosmetic and dietary angles.

Beat or blend together: 2 cups tepid water, $1\frac{1}{2}$ cups non-instant powdered skimmed milk; 3 tablespoons commercial natural yoghurt. Pour into a jug containing 1.1l (2pt) tepid water and a large can of evaporated milk. Stir well and pour into several oven-proof beakers or basins. Now place these in a large pan of warm water, bringing the water level to the rim of the containers. Cover and put into a warm oven. Keep the temperature between 37.8 and 48.9°C (100 and 120°F) for 3 hours, when the yoghurt should become thick. Store in a refrigerator. Save 3 tablespoons for making your next batch of yoghurt.

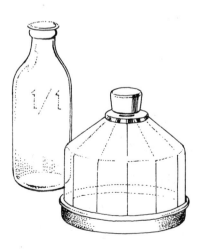

# Cleansing and Nourishing Your Skin

If you want to look good, smooth skin is a terrific asset. You can keep your skin looking smooth only if you clean it thoroughly and regularly and give it plenty of natural nourishment, especially as you get older.

## Cleansing

Every night tie back your hair and apply a cleansing cream or lotion (cream for dry and normal skins, lotion for greasy skins). The lotion doesn't have to be expensive, just generously and thoroughly used. Now massage your face and neck in circular, upwards and outwards movements, then wipe away the cream and the dirt. Repeat the whole procedure once more. This second application is necessary if you have a greasy skin with large pores, which could become even more clogged with dirt than a dry skin, or if you work in a smoky town or dirty office.

Every week, steam your face with a herbal facial sauna, unless you have very dry skin or broken veins, in which case the treatment is too drastic. Use either a tablespoon of camomile (soothing) or rosemary (stimulating) to 550ml (1pt) boiling water. Tuck your hair into a bath cap, place a towel over your head and allow the steam to penetrate the pores for about ten minutes. Pat your face dry, and allow the skin to cool before applying a natural toner such as rose-water, witch-hazel or one of the brews on pp. 8—9.

If your skin is dry, pat it with almond oil, sit in a warm atmosphere for a few minutes, then very gently remove

the oil with a tissue.

Always clean your skin very carefully around hairline, lips and the sides of the nose as pores tend to be more open in these areas and can quickly be blocked by dirt or stale make-up, allowing spots and blackheads to form.

If you have to wear a fairly heavy make-up to work, try to give your skin a breather at weekends by wearing a moisturizer only and no make-up.

In order to save money use cotton wool from a large surgical pack, rather than expensive cotton-wool balls to remove your cleanser. Tear tissues in half before you use them to remove creams or lotions. You'll find that you'll only need one or, at the most, two sheets instead of four or five. It is false economy to buy very cheap tissues or lavatory paper to remove skin cleansers as these are often hard and scratchy.

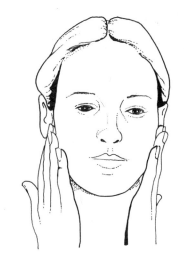

Cleansing the face

## Nourishing

Even greasy skin needs regular nourishment, and natural products are invariably the gentlest of all. For dry skin try mixing a tablespoon of milk with a teaspoon of yoghurt (natural, unsweetened). Pat the creamy substance on your face after cleansing.

For a normal skin, mash up a slice of peach with one tablespoon of cream off the milk. Apply to your skin, then rinse thoroughly with warm water.

For greasy skin, cut a tomato in half and rub the cut half all over your face. Tomato is soothing and fairly astringent without being over-drying. You can also use the pulp, mashed up with a little lemon juice and yoghurt as a facial mask. Leave it on for five minutes, then rinse.

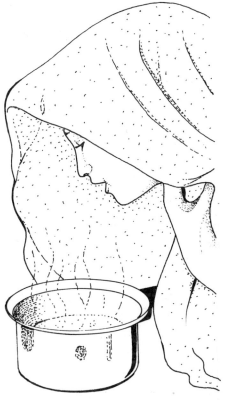

Steaming the face

# Skin Toners and Fresheners

Many fruits and vegetables contain toning and freshening properties, although some can be a little harsh and should be blended with other natural ingredients to make more gentle cosmetics. But, in general, natural ingredients do have one big advantage over commercial products—they contain no alcohol. This is often included in toning and freshening products and frequently causes dryness and soreness of the skin.

After cleansing and before applying make-up, a toner neutralizes and prepares the skin and helps close the pores. A freshener does the same job, but more gently. Toners are generally recommended for greasy skins and fresheners for normal and dry skins. But both are useful when humidity or excessive heat makes skin sweat or produce excess sebum. They then perform the valuable function of drying the skin and tightening pores. Use between make-up sessions, after cleansing or after oil-laden sunbathing stints.

Most people do need a toner at some stage in their beauty lives. Here are some excellent recipes for natural products.

### Cucumber

You can use the juice from this vegetable to make a light, natural toner and freshener. If your skin is normal or dry, it should be diluted with water. Simply squeeze the juice from two cucumbers (or liquidize them), heat to boiling point and skim away the froth. Bottle and use neat on oily skins or diluted on normal or dry skins. You can also use cucumber cut in slices. These should be rubbed directly on to the skin.

### Carrot Juice

This is good for freshening and

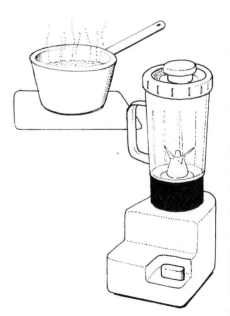

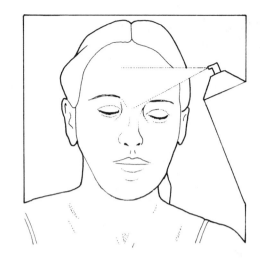

revitalizing. Use it neat, diluted with water or make it into a pre-refining mask with egg-white. Cook one large carrot gently in a little water, then liquidize or mash thoroughly. Blend with one egg-white, which has previously been whipped until stiff. Plaster this mixture over your face, allow it to dry for fifteen minutes, then rinse off.

## Onions

Although these vegetables may not smell very attractive they contain a juice which soothes and tones the skin. Try cooking chopped onion in a little water for fifteen minutes, then strain and blend the water with a squeeze of lemon juice. Use on swabs of cotton wool after cleansing and before make-up.

## Grape Juice

This is a superb natural toner. Cut the fruit in half and rub the cut half on your face. If this seems too fiddly, skin, de-pip and mash a few grapes with a little cream, yoghurt or rose-water to make a toning mask. Leave it on for ten to fifteen minutes, then rinse off.

## Lemon Juice

A good toner and freshener but it may be harsh on sensitive skins. Squeeze the juice of half a lemon into a cup, add equal quantities of water and rose-water to make a good freshening lotion. This can be used with cotton wool or in a spray to use liberally on face and neck in hot weather.

## Mineral Water

A very light, non-irritant freshener which is especially popular on the continent. Buy it by the bottle and decant it into a spray-on container. Use it to refresh your face after cleansing, to set make-up when complete, or to revitalize your complexion in hot weather. You can also use it on your neck, under your arms, between your breasts and even on your feet.

# Moisturizers and Anti-wrinkle Recipes

Finding a suitable moisturizer is one of the most time-consuming and expensive beauty chores. Many women find that some moisturizers are too greasy or too pore-clogging, or that others just don't work—in either case it's a costly business choosing! Yet every beauty expert recognizes the value of a good moisturizer as a softening, soothing agent which prepares the skin for make-up and helps prevent drying and premature ageing.

Some of the best moisturizers are natural substances such as avocado, apricots, mayonnaise, apples, cream, honey, bananas and peaches. The best anti-wrinkle treatments are oils (corn, olive, almond), cream and milk, egg-yolks and oatmeal.

If you are a dry-skin sufferer and you want to stay wrinkle-free, your diet is vitally important. You must make sure that it is rich in protein. This helps to preserve the skin's elasticity. Meat, fish, eggs, soy products, and dairy products such as cheese and milk, are all good for you. So, if you are a vegetarian, you must be absolutely sure that you get your fair share of non-animal protein— not always an easy task. It's a good idea to get into the habit of eating a salad with a vegetable-oil dressing at midday, and taking a protein-rich breakfast. Some of the leading, older beauties in the world have rejected total vegetarianism in favour of a more relaxed diet which allows eggs, milk and fish—all valuable forms of protein.

## Moisturizing Recipes

### Avocado Face Pack

Mash softened avocado with a little milk or cream to make a paste. Plaster on to a clean skin, rub in and leave for fifteen minutes. Rinse off with tepid water.

### Apricot Mask

Cook three or four dried apricots in a little water until softened. Mash with a teaspoon of honey and a few drops of cider vinegar. Blend with enough milk to make a fairly creamy substance. Pat on to the face, leave for twenty to thirty minutes, then rinse off.

### Mayonnaise Treatment Cream

Make and bottle mayonnaise in the normal way, using one egg to 140ml ($\frac{1}{2}$pt) olive oil, but omit the seasoning. For a soft, smooth skin, pat on to the skin daily, leaving for five to ten minutes, then rinse off with tepid water.

### Apple Juice Moisturizing Tonic

You can use home-blended apple juice (add water to chopped apple and blend) as a daily moisturizing tonic. It should be left on the skin then rinsed off. Alternatively, you can cook and mash eating or cooking apples, adding a little honey and cream to make a super moisturizing mask. The pectin in the apple and the softening properties of the cream and honey combine to do wonders for your skin. This is a cheap and easy beauty treatment to use if you're preparing apple purée for the freezer or to go with roast pork!

## Banana Nourishing Mask

Bananas are rich in Vitamin A, and contain a natural, soothing oil. Try rubbing a used banana skin (inside out) on your face to see how soft and soothing it feels. Use an over-ripe banana like this : mash the fruit, add a spoon of cream, milk or yoghurt and a little honey (also add an egg-yolk if your skin is very dry). Blend the mixture and plaster it on your face, then rinse off with tepid water fifteen minutes later.

## Anti-Wrinkle Treatments

Laughter lines can be attractive, but frown and misery lines definitely aren't. Try to avoid frowning too much when you're working, driving or watching television. Take a break to raise and lower your eyebrows a few times when you know that you're frowning too much—and don't let the corners of your mouth droop either. But if you know that you are likely to develop unattractive lines on your face, try to minimize the effect by using extra moisturizing treatments on these areas. Every night rub a little almond oil into each wrinkle, using a gentle touch (a small bottle of the oil should last for ages). This will help to prevent the wrinkle deepening. Try the following too :

## Egg and Oatmeal

Make up one tablespoon of porridge oats with water or milk to a fairly thin consistency. Blend in one beaten egg. Plaster the mixture over your wrinkles and leave on for fifteen minutes, then rinse off.

## Milk and Honey

Warm half a cup of milk, stir in two teaspoons of honey. With a little cotton wool pat the mixture gently on to wrinkles. Leave for thirty minutes, then rinse off with tepid water.

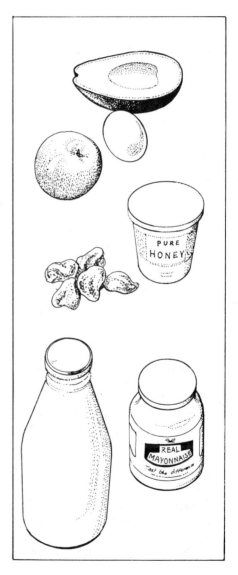

# Greasy Skin and Acne

Excess sebum flowing from the pores of the skin causes spots and pimples when it mixes with dirt and debris to block the pores. The biological malfunction that makes the body produce excess sebum in the first place is, like so many beauty problems, hormonal in origin. Androgen, a male hormone present in females too, becomes over-active during puberty and one of the side effects of this is excess sebum. This is why greasy skin is usually a big problem in your teens and early twenties and is more pronounced in boys than girls.

If you have an excessively greasy skin, spots and blackheads, then it may help if you watch your diet. Avoid fried or spicy foods, oily dressings and too much carbohydrate. Eat fresh citrus fruits (Vitamin C is a valuable skin food), and drink lots of water.

You should also aim to keep the greasy secretions under control with twice-daily washing using a bland soap or special washing grains (see recipe below). Make-up is a good idea as worrying about spots can make the hormonal activity even more pronounced, so disguising them can actually help clear them up. But use a very light make-up and don't buy commercial products which over-dry the skin, otherwise your face may become itchy, painful and even swollen on the outside, with those glands still pumping away there on the inside. A hypo-allergenic liquid make-up is a good idea since a greasy skin is often a sensitive one, as well—not the tough covering it is popularly thought to be. That's why sunshine, which can certainly help dry up grease and make the skin smoother, should only be used in moderation as a free beauty treatment. Try sunbathing for just a few minutes every day ; even if it's quite cool, leave off your make-up and sit in the garden wearing a woolly jumper. You'll find your spots will improve a lot. Start the treatment in April or May when the sun can be powerful enough for a fresh-air facial.

## Washing Grains

Mix a little cold water with a handful of ordinary porridge oats and rub the mixture on to your face, paying particular attention to blackheads and spots. Rinse off with tepid water. Oats are soothing and cleansing and, unlike perfumed soaps, leave your skin in a bland condition.

## Melon

A super fruit for oily-skin masks as it is moderately astringent yet very gentle, needing gentle treatment. Mash a tablespoon of soft melon flesh with a little home-made or shop-bought natural yoghurt. Beat up an egg-white, and fold it into the mixture. Plaster it on to your face, allow to dry for fifteen minutes, then rinse off with tepid water.

## Yoghurt

This is helpful in controlling excess grease. You can use it combined with any of the following ingredients to make a mask : mashed cucumber, mashed potato, mashed tomatoes or tomato purée, lemon juice, or ground almonds with a dash of rose-water added. Plaster the mask on to your face, leave for fifteen to twenty minutes, then rinse off. I advise using yoghurt as the binding substance as it is gentle but if you have very greasy skin, you could also add stronger, more astringent binders, such as fuller's earth, kaolin or, perhaps, a few drops of Friar's Balsam.

## Lemon

This is certainly astringent, and should be used with care as it is so acidic. Lemon juice mixed with a little warm milk or yoghurt makes a good skin lotion. Swab on to the face with cotton wool, allow to dry, then rinse off with tepid water. Alternatively you can squeeze the juice of half a lemon into a cup, add a beaten egg-white and a few drops of rose-water to make a skin-tightening lotion which is useful if your skin is looking muddy and greasy and you want to revive it quickly before a special occasion. Pat the mixture on to clean skin, lie down for thirty minutes, then rinse off with tepid water. You'll find that your face instantly looks softer, clearer and less spotty.

## Blackheads

It's hard to resist squeezing blackheads if you have a lot of them, but do remember that doing this with dirty fingers will probably make your face more spotty than before. If you have a few minutes' privacy then try getting rid of a few of the demons like this: clean your face, and wash with bland soap. Heat up half a cup of milk until lukewarm. Smooth over the skin with cotton wool. Wait for a few moments to allow the milk to penetrate the pores and soften the blackheads—you can use the time to wash your hands very thoroughly. Now squeeze out the blackheads gently using those well-scrubbed forefingers and try not to dig in your nails. Finally, wash your face gently in tepid water.

## Open Pores

Astringents will only close pores temporarily by irritating them slightly and causing the skin to contract around them; they open again afterwards. However, open pores are funnel-shaped; the wide part of the funnel is usually made up of skin debris, dirt and stale make-up. If you can clean it all away regularly, the pores will look smaller. Ground almonds blended with rose-water, milk or yoghurt are good for this purpose. Make a cleansing mask and use it once or twice a week. Washing the area with a sage infusion (half a cup of leaves to 225ml ($\frac{1}{2}$pt) boiling water, steeped for ten minutes, then strained), will also help. Elderflower water (see stockist list on p 5) is a good cleanser, and even more effective if you add a squeeze of lemon juice. Apply this on cotton-wool.

Many people with greasy skins also have very dry areas around their eyes. Do *not* spread any of the above treatments over these dry areas.

# Soft, Beautiful Lips

It's fashionable to have soft, shiny lips but the skin covering the lips is so delicate that it often becomes chapped and dry. Applying lipstick to dry lips produces a bitty, flaky result, and adding gloss on top creates a mess. Try these ideas to give your lips luscious looks.

## Lip Balm

Saliva is the cheapest of all. It contains a natural antiseptic, soothing ingredient (the reason why animals lick their wounds) which softens and heals the delicate tissue. Lick your lips after removing lipstick at night and when you wake up. For more serious chapping, rub another natural soother, honey, on to your lips. Yoghurt is also good and helps get rid of mouth troubles like ulcers and sores.

## Glossers

To protect and beautify your lips, using a lip-brush, simply apply a dab of petroleum jelly such as Vaseline. This will add gloss, too, and looks good alone or over lip colour applied in pencil- or stick-form. Victorian ladies used to bite their lips to give them colour—this works, but is a bit painful. Add a drop of peppermint, rum or almond essence to the pot of petroleum jelly if you want a flavoured effect. Baby or olive oil can also be painted on to the mouth with a soft brush.

## Tints

Food colourants and additives, such as cochineal and milk-shake powders (strawberry or raspberry flavours), can be blended with baby oil or petroleum jelly to make interesting and tasty lip-tints. Also try warming up all those old, unfinished lipstick stubs together with a little lanolin or olive oil to make soft lip paint. Pour it into a little pot such as an empty eye-shadow case or a child's paint palette, and apply with a brush. Lip-brushes are invaluable for the correct application of lip colour and for saving money. Use a paint-brush, or wash an eye-shadow brush when the shadow is finished, for your lip-brush. For an even result, top make-up artists always apply foundation over the lips before making them up. If you feel your normal foundation is a bit harsh for your lips, mix it with some moisturizer on the palm of your hand, before rubbing it into the lip area.

## Tissues

Brighter lip colours have brought the blotting process, by which surplus colour is blotted on to a special tissue after application, back into fashion. Ordinary face tissue tends to absorb too much lip colour but the shiny, hard, institutional lavatory-paper is ideal. Cut it into small pieces and tuck them into your make-up bag ; their shine will leave a smooth surface on your lips.

## Lip Exercises

Say 'O' and 'E' rapidly to give your mouth a fuller shape. Then make up a sentence with as many words as possible beginning with the letter 'W' (eg 'Why does wet weather wait until Wendy's wellingtons are too worn to walk to Worthing in ?)'. Now whistle.

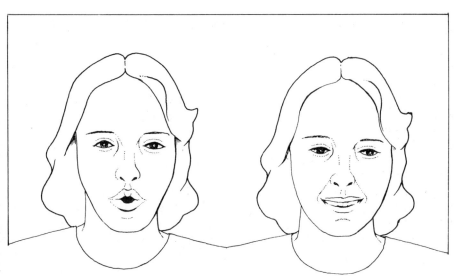

# Sparkling, Beautiful Eyes

Sparkling eyes are yours for free with the right care. Eyes are so important in today's ideal of beauty that it really does pay dividends to look after them. Sensible eye-care is good also for your general health and well-being.

Firstly, make sure that your diet is high in Vitamin A, found in carrots (try grated carrot on salads), apricots, liver and turnips. This is the top vitamin for eye health. Drink water, avoid too much alcohol and try to give up smoking to keep your eyes clear and sparkling.

Secondly, don't strain your eyes unnecessarily. Never watch television in a darkened room, and always look slightly down on to the set. When working make sure the work itself is well lit. Wear glasses if you have to and, if in doubt, get your eyes tested. If you have a long session at your desk or in front of a typewriter, break off now and again for some eye exercises. Sleep is the best eye-brightener of all.

Thirdly, be choosy about eye cosmetics. Use the natural remedies below where possible and, if your eyes are sensitive, choose hypo-allergenic commercial cosmetics. Apply eye-liner with care and don't let particles of pencil or kohl drop inside the eye. If you use nail polish and your eyes are inflamed, your polish could be the villain. For beautiful eyes try these natural treatments.

## Tired Eyes

Lie down for fifteen minutes with a used, cold teabag or a slice of cucumber on each eye. The tannin in the teabag or the cooling juice of the cucumber will soothe and brighten your eyes.

## Puffy Eyes

Step up your water intake, try and manage eight glasses a day. Grate half a large potato, divide in half and wrap each half in muslin or thin cotton to make a pad. Place the pads over the swelling and rest for fifteen to twenty minutes. Alternatively, you can use cotton pads soaked in witch-hazel or cold camomile tea, or a fresh fig cut in half and placed over the area.

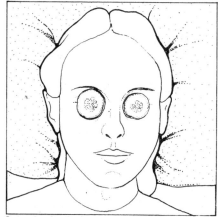

Cucumber for tired eyes

Applying almond oil to eyebrows

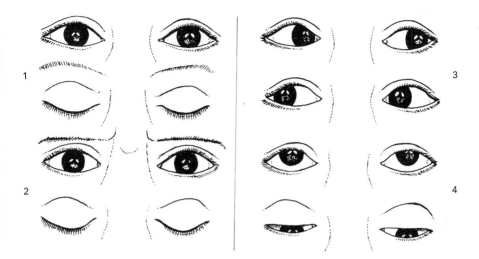

## Bloodshot Eyes

Eyebright is a herb traditionally associated with eye treatments. It grows wild on chalky soils, and has white or purplish flowers. Use 28g (1 oz) dried eyebright to 550ml (1 pt) boiling water to make an infusion. Strain, cool and bathe your eyes with the lotion three times a day.

## Sparse Lashes

Help encourage growth by brushing them with a little almond oil on an eyebrow-brush each night. Use only liquid or cake mascara, long-lasting mascaras may be difficult to remove and cause the lashes to become even more sparse.

## Wrinkles

The skin around the eyes is the most delicate area of the face, so never use a heavy lotion or cream on this area. Instead, pat a few drops of almond oil on to the skin at night. When available, pat any left-over water-melon juice around the area (having eaten the flesh first). This contains a natural skin-softening enzyme which will help tighten slack, wrinkly skin.

## Straggly Brows

Use almond oil on a brush to smooth brows into a neat line and to soften them before plucking—this will make eyebrow-shaping less painful.

## Eye Exercises

1 Look straight ahead, eyes wide open and fix your gaze on a picture or wallpaper pattern on the opposite wall. Raise your eyebrows and close your eyes tightly. Rest, repeat five times.
2 Open your eyes look straight ahead, frown slightly and close your eyes tightly. Rest, repeat five times.
3 Look as far to the left as you can, then as far to the right without moving your head. Repeat ten times.
4 Raise your eyes to the ceiling, then lower them to look at the floor, again without moving your head. Repeat ten times.

# Complexion

If you really want to save money, don't buy powder, blusher or highlighters for there are lots of other cosmetics already in your beauty bag which can do the same jobs. If money is limited invest in good brushes to shape your complexion : a large, soft brush for powder, a smaller, soft one for blushers and a whole clutch of little brushes for eye-shadows. Look in the art department of your local store for hog's hair or camel-hair brushes or buy cosmetics brushes from your local chemist. Keep them clean with regular washing. It's interesting that the most professional-looking make-ups are always applied with brushes yet the standard brushes supplied with most blushers and eye-shadows are very bad indeed. So, if you do go out and buy a blusher or highlighter, look for the refill of a cheaper, brushless pack and use your own good brush !

## Powders

A popular look just now is pearly, a soft, beige complexion topped with a translucent powder. Achieve this look with a good liquid foundation in a beige shade topped with baby powder. Although baby powder looks dead-white in the pack, if you dust it on to your face with cotton wool, then brush lightly with a baby brush the effect is pearly and translucent—just as good as an expensive powder product. You can also use beige or rose-tinted talc for a warmer effect, or you can add a tiny bit of powdered brown or rose-coloured chalk to basic baby-powder. Scrape the chalk very gently with a razor-blade, then mix it together with the powder.

## Highlighters

These are used to add highlights to your face giving prominence to those bones at the temples, your cheek-bones, chin, etc, which make your face-shape more interesting. Remember that using a highlighter on a certain part of your face will bring that part forward, so, if you have super cheek-bones, they are obviously suitable areas for using a highlighter. Blushers either make the area recede (dark shades) or add warmth (rosy shades).

Try using these cosmetics as highlighters : pearly beige eye-shadow, pinky cream or stick eye-shadow, yellowy-beige eye-shadow, clear lip-gloss, baby powder, pale, pearly-pink lipstick.

## Blushers

Use a blusher to make an area of your face recede, that too-wide jawline, those chubby cheeks or the sides of a very full forehead. Choose browny-beige shades for this job ; browny-beige

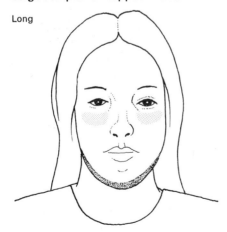

Long

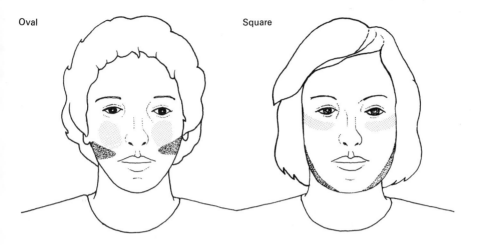

Oval

Square

eye-shadow is perfect, as are brown eye-shadows (powder) or tan face powders blended with baby powder. Cinnamon-brown lipstick mixed with moisturizer and applied with a delicate touch then topped with powder will also do the trick.

Use a blusher to add warmth to your face and put a sparkle in your eyes, too. Smooth it on above your cheek-bones, outwards and upwards in a semi-circle towards your ears. Pinky-beige lipstick and lip-pencil are both excellent for this as are tinted lip-gloss, Vaseline mixed with a dash of cinnamon lipstick, pinky-brown eye-shadows (cream or powder types) and pink-tinted talcs.

See the illustrations for disguising tricks by using highlighters and blushers for different face-shapes. Highlights should be used to bring out features which need emphasis, dark shadows and blushers to fade out poor or over-large features. Blend all the colours very subtly into the rest of your make-up and top with powder to make the effect even more subtle.

Round

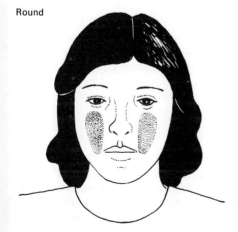

(*opposite page and this page*)
Using highlighters and blushers to emphasise or disguise features

# Beauty and the Bath

The modern bathtime ritual is a useful anti-stress routine, but it is often very bad for the skin. Many bath preparations are harsh, stripping the natural, softening oils from the skin and cleaning away friendly bacteria which actually protect us from skin troubles. Indeed, many dermatologists think daily bathing is bad for hygiene.

Natural ingredients—oils, herbs and vinegar, particularly—can soften and delicately scent your bath-water inexpensively, while taking the harshness out of a highly relaxing beauty routine. Try these ideas:

## Pre-bath

Rub your body with coarse sea salt, or

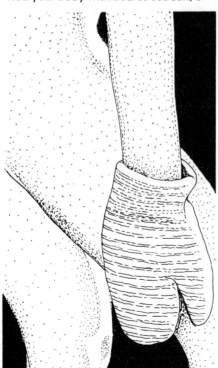

epsom salts, if your skin is dull and grey-looking, or if you have the yellowish remnants of a tan.

Try a rub down with a cider vinegar and water mixture (two parts water, one part vinegar) if your body is hot and itchy after a long day at work or after travelling in crowded public transport.

Use a cupful of porridge oats tied in a cotton hanky if your skin is sore and dry. Just rub the cotton porridge pad all over your body several times before stepping into the bath.

Soften dry, tender skin, or prolong a tan, by rubbing yourself all over with corn oil and standing in a warm spot for a few minutes before slipping into the tub. This will allow the oil to soak into the skin and soften it.

Stimulate sluggish circulation with a dry rub, using a bath mitt, loofah or natural sponge. This helps to slough off dead skin-cells and make your skin tingle.

## Bath Additives

Oil is the best ingredient in your bath if you want your skin to stay young. Add a capful of baby or corn oil to your bath, with a very small amount of aromatic perfume essence to take away the oily smell. The oils disperse over the water, clinging lightly to your body as you emerge from the tub.

Milk does much the same job as oil. Add half a cup to your bath but do include some perfume or perfume essence as the milky smell isn't very attractive.

Herbs like camomile, rosemary, blackberry leaves and basil will add a soothing quality to the water and give a heady, sensuous smell. Either pre-soak a tablespoon of mixed

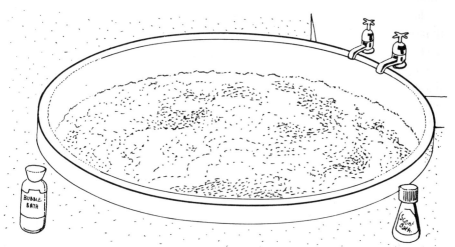

herbs or of your particular favourite in a cup of boiling water for fifteen minutes, then add the strained liquid to the bath, or tie the herbs in a cotton hanky or muslin bag and place it under the taps while the bath is running.

Sun-oil can be a great bath additive as it gives off a heady aroma as well as softening the skin. Use up leftover summer suntan oils by adding them, a capful at a time, to your bath.

Perfumes for bathtime use should be sensuous. If you have musk oil, geranium oil or a bottle of oil of cloves or oil of peppermint you can make up your own scented bath oil by mixing them or using them on their own, a drop at a time. Peppermint oil is particularly refreshing, and clove oil smells extremely sexy.

Soap should be bland and gentle, so avoid expensive, highly perfumed, frothy soaps. Baby soap is ideal. It's possible to make your own soap but it is rather an involved process. For cleansing without soap, use a handful of oatmeal or a paste of bran and water; or, mix up a brew of baby shampoo, olive or corn oil and a little perfumed oil. Shake and lather your body with the mixture before tubbing.

## After Your Bath

Don't be harsh with your skin. A towelling robe will dry your body gently, or try wrapping yourself in a soft, fluffy towel and pat your body all over. Smooth more baby oil into your skin, working it well between your toes and into knees and elbows.

Try to relax for a few minutes before dressing, to allow your body temperature to return to normal, otherwise your clothes will feel itchy. If you can, walk about naked for a little while to feel the wonderful freedom of truly natural beauty !

## Shower Power

Try stimulating circulation with a warm shower which, very gradually, becomes an icy cold one ! Just adjust the taps slowly until you feel tingly. Remember that the most economical way of using your shower is to wash head and shoulders first, working down to your toes.

# Pretty Feet

Keep your feet healthy and attractive by wearing comfortable shoes. In winter, choose boots with wide toes and medium-height heels and shoes that grip across the instep and around the ankle. Sandals are fine in summer as long as the heels are sensible and the straps don't rub. Go barefoot wherever and whenever it's safe to do so.

At the end of a busy day, dabble your toes in lukewarm water with a tablespoon of cider vinegar added to stop itchiness and tiredness, and freshen up your feet. If you often suffer from tired legs and cramps, bolster the

Vitamin C in your diet by eating two oranges a day or sipping natural lemon juice in hot water every night.

For really rough skin, soak your feet in salted water, dry and then rub them all over with corn or olive oil. Blot with tissues before putting your shoes or tights back on. If you have to do a lot of walking, a daily oil rub can make a fantastic difference to the comfort and appearance of your feet.

Try this weekly step-by-step pedicure treatment—most of the equipment can be found in your store-cupboard apart from the scissors, pumice stone, orange sticks and water.

## Step by Step Pedicure

*You need* One tablespoon of salt, olive or corn oil, pumice stone, two tablespoons of cider vinegar, lemon peel, nail scissors, orange sticks, French chalk or henna powder, a bowl of warm water, nail buffer or piece of chamois leather.

1  Remove tights and shoes and spread out an old towel on the floor. Add the salt to the water and dabble your feet in it for about ten minutes.

2  Dry your feet thoroughly, and massage well with the olive or corn oil paying particular attention to rough areas (soles of feet, ankles, etc) and any callouses. Work oil well between the toes.

3  Allow the oil to soak in for a few minutes. Rub callouses gently with the pumice—the softened skin should be easy to smooth away.

4  Throw away the salt water, add fresh warm water to the bowl, plus the cider vinegar. Place your feet in the bowl, and rinse thoroughly.

5  Place each foot in turn on a tissue on the bathroom chair and deal with those toe-nails. Push back the cuticles with an orange stick and rub thoroughly with the pith side of the lemon peel. Repeat with all ten toes.

6  Cut your nails carefully, straight across (this is important). Use the orange stick to clean behind the nails. Dabble your feet in the vinegar-water again.

7  Allow your feet to dry naturally. Place a spot of French chalk on each nail and buff to a super shine. Alternatively, mix henna powder and a little water to make a paste and paint the mixture on your nails with a child's paint-brush. Allow to dry, then rinse for a very subtle orangy-pink shine. Now you should be walking on air!

**Treats for Your Feet**

Put your feet up—right up the wall, in fact. Lie with your bottom against a wall and your feet straight up the wall in front of you for five minutes. This will put the spring back in your step.

If you have elderberries in your garden, dry the leaves, crush them and add a little fuller's earth. The powdery substance can be sprinkled in your shoes just before a shopping expedition to help prevent sore feet.

Rotate your feet from the ankles, left and then right. You can do this when you're sitting down in the office, or watching television. This will help strengthen your ankles and your feet.

Soothe swollen feet and ankles by tying a crushed, dampened geranium leaf over the swelling and putting your feet up on a pillow or footstool for a while.

# Beautiful Hands

Save money on handcare by making sure you look after your hands and nails carefully. Wear rubber gloves for housework and washing-up, particularly when using harsh household cleaners and scouring powders.

Choose bland soaps for your bathroom and kitchen. Always keep a piece of lemon peel by the sink (left over from your cooking). Use it to rub into hands, around nails and elbows to soften and strengthen the skin.

If you have to do a really rough job—something fiddly with the car for instance—and can't wear gloves, use this natural barrier-cream : mix two teaspoons of fuller's earth, two teaspoons of olive oil and one egg-yolk. Rub thoroughly into your hands and under your nails. After work, rinse your hands carefully in lukewarm water and rub in a little more oil.

Never throw away cold porridge. Use a porridge and water or porridge and milk paste on any rough patches of skin, on arms, elbows, ankles or face. Plaster the mixture on, rub in gently and leave for fifteen minutes, then rinse well.

## Step by Step Manicure

*You need* Lemon peel, cider vinegar, camomile (dried or fresh) or marigold petals, egg-yolk, honey and salt, French chalk or henna powder, orange sticks, emery boards, a small bowl of warm water and a nail buffer or a piece of chamois leather.

1 If you are wearing nail polish, take it off carefully with cotton wool and an oily remover, working all over the nails in small circles until all traces of it have vanished.

2 Rub all over your nails and hands with the pith side of the lemon peel. Rub under the nails and into the cuticles pressing them back gently. This will strengthen and whiten them and make them smell good too.

3 Shape your nails with an emery board—make them rounded, not pointed, and take care to file from sides to centre in long, gentle strokes.

4 Add two teaspoons of cider vinegar to your bowl of water and dabble your fingertips in it for five minutes. This will help strengthen brittle nails. If your fingers are very chapped and your cuticles rough and torn, try this remedy : add a tablespoon of dried camomile or marigold petals to a cup of boiling water. Allow to infuse for fifteen minutes. Strain the liquid and soak a soft hanky or piece of muslin in it. Wrap the hanky around one hand and leave for six minutes, then resoak it and apply to the other hand.

5 Mix the egg-yolk, one teaspoon of honey and a pinch of salt. Work well into your nails and cuticles to nourish and strengthen them. Leave for five minutes, then rinse.

6 Clean underneath your nails with the orange sticks, carefully removing traces of the strengthening egg-yolk and honey mixture. Rinse your hands in the vinegar and water to clean thoroughly.

7 Allow your nails to dry thoroughly (don't use a towel). Place a little French chalk on each nail and buff with the nail buffer or chamois leather until it shines. For a long-lasting fashionable orangy-pink nail enamel, mix henna powder with warm water to

make a paste and paint it on to the nails with a child's paint-brush. Allow to dry, then rinse the paste off in lukewarm water.

## Natural Hand Lotions

Carefully heat three tablespoons of glycerine in a pan. Mix two tablespoons of cornflour with a little of the glycerine, returning it to the pan to thicken. Add half a cup of rose-water. Cool and bottle. Melt 14g ($\frac{1}{2}$oz) white wax with six tablespoons of corn oil and one teaspoon of cod liver oil. Now beat in 56g (2oz) rose-water very carefully, drop by drop, to blend thoroughly with the oily mixture. Use a fork for this process. Pour into jars and cool.

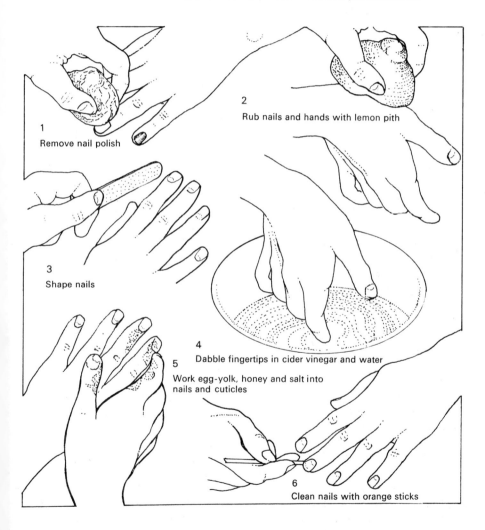

1
Remove nail polish

2
Rub nails and hands with lemon pith

3
Shape nails

4
Dabble fingertips in cider vinegar and water

5
Work egg-yolk, honey and salt into nails and cuticles

6
Clean nails with orange sticks

# Your Crowning Glory

Hair care and condition is currently receiving wide attention. Hairdressers are promoting natural treatments and colouring processes which date back centuries. You can give yourself the same treatments at home for a fraction of the cost and get superb results.

If you have dry hair, you should first understand what is causing the problem. The natural conditioner which imparts sheen and gloss to the hair shaft is called sebum. It is secreted by the sebaceous glands in the hair follicles all over your head ; these house the root of each hair. If the sebum is not produced in sufficient quantities, it will only condition the top of the hair shaft, leaving the ends feeling dry. If you have very long hair, the problem may be aggravated. To help treat the problem, first make sure that your diet contains the correct foods to enable your body to produce sebum. Natural vegetable oils, yeast, wheat-germ, liver and kidneys can all help. Trichologists usually prescribe about four to six yeast tablets a day for dry hair. You can also drink stout, which is a good natural source of Vitamin B.

Give your scalp a daily massage to stimulate the circulation and encourage sebum to flow freely. Place the tips of your fingers on your scalp and make circular movements. Make sure you are moving your scalp, not your fingers, in the process.

You should also have your hair trimmed so that the glands have a fighting chance of catering for the whole length of the shaft. It's unreasonable to expect glossy hair if the ends of your hair are three years old

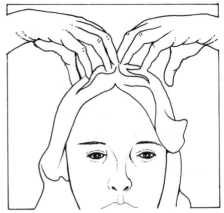

Massaging scalp

—and, at a growth rate of about half an inch a month, the ends of your long hair can easily be that old !

Try one of these pre-shampoo treatments :

## Honey and Oil

Mix in a cup one part honey with three parts of olive oil (use a larger quantity for very long hair). Leave the mixture to sit for a day. Stir well and apply to scalp and hair before shampooing. Put some plastic cling-film on your head to stop the drips and relax for half an hour. Now shampoo twice, rubbing well, and add cider vinegar or lemon juice to the final rinsing water.

## Egg and Oil

Beat two eggs in a bowl (one if your hair is short), then whisk in one tablespoon of corn or olive oil, one tablespoon of glycerine (optional) and one tablespoon of cider vinegar. Use as above.

## Hot Oil

Warm a cup of olive, castor or corn oil. Use as above, but put a warm towel on top of the plastic cling-film or sit in the sunshine to give additional warmth

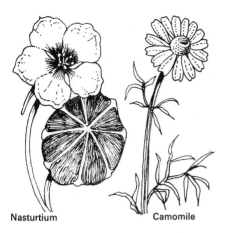

Nasturtium      Camomile

while the oil penetrates the hair shaft.

### Oil and Rosemary

Put a tablespoon of dried rosemary leaves into a small, glass, screw-top jar containing a cup of cooking oil. Close tightly and leave in the sunshine or an airing cupboard for two days. Strain and use as above.

After washing your hair with a mild shampoo, comb it through carefully and rinse with an infusion of rosemary (dark hair) or camomile (fair hair) to give lustre and manageability. Simply make up a strong infusion of either herb in a teacup and strain straight into your final rinsing water.

## Dandruff

This is usually associated with dry hair, although greasy hair can also be afflicted with this scaly condition at the roots. In adolescents, excessive oiliness caused by androgen hormones triggers off the infection. But other causes can be emotional, or the result of a faulty diet. To help eliminate dandruff avoid foods which aggravate the problem, such as animal fats, fried foods and chocolate.

Then brush the hair daily before dressing with a clean bristle brush to get rid of most of the tell-tale white flakes. Next, and most important, try one of these between-shampoo scalp treatments. All should be applied to dry hair, rubbed well into the scalp and allowed to dry thoroughly before the hair is styled or brushed.

### Nasturtium

Gather a cup of nasturtium leaves, tear and bruise them, then pour on boiling water. Allow to steep for fifteen minutes. Strain, cool and apply to the scalp on cotton wool. Rub in well.

### Nettle

Gather about two tablespoons of young nettle leaves (wear gloves), tear roughly and steep in 225ml ($\frac{1}{2}$pt) boiling water. Strain and add a tablespoon of cider vinegar. Massage as above.

### Witch-Hazel

Mix half a cup of witch-hazel with a squeeze of lemon juice and a tablespoon of cider vinegar. Massage as above.

# Greasy Hair

Many people, especially teenagers, suffer from greasy hair and an oily scalp. The cause is over-active sebaceous glands triggered off by hormonal activity, especially prevalent in puberty, particularly for young men. This activity can be regulated to a certain extent by careful diet control. Drink plenty of water, eliminate fatty, fried and spicy foods, and increase your intake of salads and fish. If you can, drink lemon juice and water each morning.

When your hair becomes greasy, you should wash it carefully, using a mild shampoo—it is not true that over-washing aggravates the sebaceous glands as long as you wash the hair and do not rub the scalp itself too hard. Many people shampoo their hair wrongly, rubbing the scalp and ignoring the greasy ends. Always apply the shampoo to your hands, not directly on to the scalp, then work it into the hair and rinse from the roots downwards, so that dirt, grease and debris flow into the sink or bathtub. Try adding a cupful of rosemary infusion to your shampoo—one cup of boiling water to a tablespoon of rosemary leaves, cooled and strained.

After washing, rinse your hair in cool water to which you have added a tablespoon of cider vinegar or a squeeze of lemon juice. This will eliminate the soap and restore a good, protective, acid mantle to the hair.

Between shampooing there is a lot you can do to keep greasy hair looking fresh. Try brushing with a natural bristle brush, covered with a fine cotton hanky or piece of muslin, dampened with eau de cologne. This will absorb grease and freshen the hair. You can also make your own dry shampoo as follows:

### Dry Shampoo

Mix one part fuller's earth, with two parts talcum powder. Dust on to the hair, leave for a few minutes, then brush out. This is strong-acting, so don't leave it on overnight. Do make sure that you brush out the powder thoroughly.

### Scalp Rub

To help staunch the flow of sebum, try rubbing your scalp with cotton wool soaked in witch-hazel, rose-water or eau de cologne. Simply make partings all over your head, rub the parting with a lotion-soaked piece of cotton wool. Allow to dry, then comb into shape.

### Greasy Hair Dandruff

This can form on your scalp,

Applying dry shampoo

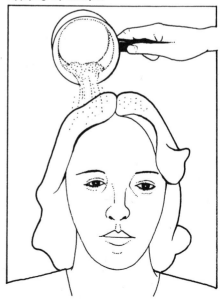

encrusting the hair root. One of the above remedies will help it, as will cider vinegar or nasturtium lotion (see p 27).

## Controlling Your Body

If you are past puberty, and lead a hectic life, then stress could be causing your greasy-hair problem. Follow the dietary advice above and watch out for the possibility of poor thyroid function. Increase your intake of fish and iodine-rich sea salt. Relax your whole system with yoga-based exercises every night.

## Body Control Exercises

1 Lie on your tummy, hands palm downwards under your chin. Breathe deeply. Now place your hands under your shoulders and stretch your arms so your whole torso is lifted off the ground. Push your head back and look at the ceiling. Hold briefly, then relax.
2 Lie on your back, hands under your

Rubbing scalp with rose-water or witch-hazel

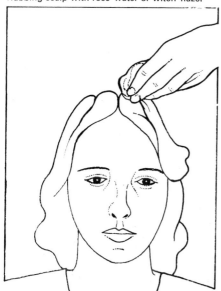

bottom. Now raise your legs, supporting your bottom with your hands. Pause, then raise them further until you are 'standing' on your shoulders with your hands supporting the small of your back. Hold, then relax.
3 Lie flat, breathing deeply, hands by your sides, palms uppermost. Close your eyes and concentrate on your slow, deep breathing. Feel your whole body become heavy and limp. Stay in this position for ten minutes.

## Care and Hygiene

If you have greasy hair, do make sure that you wash your brush and comb every time you shampoo your head. Keep them clean in a drawer or cotton bag, away from dust and bacteria.

It's more sensible to wear your hair off your face or in a short style, since grease transferred from the hair on to the face can cause spots to build up— over-active glands on the scalp usually go together with over-active glands on the face.

When removing make-up, pay particular attention to the area along the hairline. The accumulation of make-up and dirt will aggravate your hair problem. Fasten your hair well back from your face, so you can see exactly where you are cleaning. If you use a nourishing night cream, make sure it does not touch your hairline. Beware of wearing synthetic scarves or woolly hats which may cover the problem temporarily but certainly make it worse. However, if you work in a dirty, grease-laden atmosphere (in a restaurant or factory, for instance), a very light cotton scarf would be a good idea.

# Setting and Styling Your Hair

If you can set your hair at home yourself, you will save pounds on hairdressers' bills. Although there is no substitute for a really good professional cut or perm, a home hair-do in between times can be chic and attractive.

Don't spend pounds on shampoos, you will often be paying as much for the packaging, froth and smell as the cleansing-power. Choose a mild, inexpensive baby shampoo even if your hair is greasy.

## Setting Lotions

Try adding a cube of ordinary-flavoured jelly to the final, warm rinsing water—lemon for fair hair, raspberry or strawberry for redheads or brunettes. This will give body to the hair and make it more manageable. If you have fair hair, use lemon juice and water as your setting lotion (juice of one lemon to two cups of water), combing through each tress of hair before you put it in a roller, or combing through the whole head before blow-drying.

## Setting and Styling

Nowadays, most styles simply need blowing dry with a good brush and dryer, but some styles do still need the added bounce of rollers.

## Roller-setting

Towel-dry or allow the hair to dry slightly before setting. It takes longer, and wastes electricity to dry hair from soaking wet to bone-dry. Use cheap non-spiky plastic rollers—bristles, spikes and metal are all bad for the hair. Part the hair, and take small sections for each roller, starting with the hair at the crown of the head. Hold each section at right-angles to the head as you wind, to make sure you get maximum bounce, and comb it through before winding. Back, nape and side rollers should be small. Use plastic or covered hair-pins to avoid damaging your scalp. Dry in the fresh air in summer and use a hooded hair-dryer in winter to give even air distribution and prevent arm-ache.

When dry, first brush the hair in the opposite direction to the finished style to give volume. Style carefully, without resorting to damaging back-combing.

## Blow-Drying

This takes practice, but is easy when you know how. Comb the hair through carefully first and let it dry off slightly before you begin. Grasp the dryer in your left hand and the brush (which should be firm and rounded in shape) in your right (or the other way around if you are left-handed). Now, holding the dryer about 15cm (6in) away from your hair, lift the hair in sections with the brush, using either under or upwards movements according to the style required. Start with the top and back hair, end with the front hair as your arms will be most tired at the end. You may find that you need to dampen your fringe and front hair a little to complete the style. You should always lift the hair up away from the roots to give bounce. If your hair is long, you may need to pin the rest of the hair out of the way, with a long clip or two, while you work on each section.

## Using Tongs

If you have a short hair-do, then curling tongs can be a good investment. Start with the side hair, placing a comb underneath the tongs to prevent burning. Make twirled sausage-shapes all over the head, then allow them to cool. Brush away from the direction of the finished style, then comb or brush into shape.

## Trimming Your Hair

If you have a fringe or a short hair-do, save money by learning to trim the straggly bits yourself. Always cut hair wet, using sharp scissors. For your fringe or side-curls, comb the wet hair on to your face and secure with thick Scotch tape (not the transparent kind) before cutting carefully. Remember that your fringe will bounce a little shorter when it's dry.

## Refreshing Your Set

Spray your hair with a fine spray of water to which you've added a little eau de cologne, then pin in rollers, take a bath and let the steam help set the style. Afterwards, dry with a hand-dryer or sit in a warm room for a while. If the style goes floppy or greasy, brush through even quantities of talc and fuller's earth to restore the style.

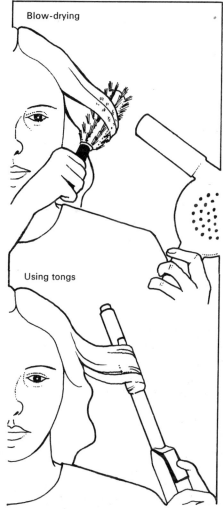

Blow-drying

Using tongs

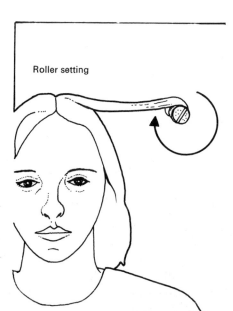

Roller setting

# Hair Colour and Shine

The most successful hair colour is the colour that looks natural. If you decide to change your hair colour, always choose a shade that is near your natural tone. Radical changes look hard and are expensive to keep up since new growth shows. Always make sure that you and your hair are in good health before you use a colourant. Hair is sensitive and tends to react differently to chemicals or natural ingredients according to one's body chemistry. If you're not entirely fit and well, your colour change could be less successful than you'd hoped. So, top up your diet with B vitamins (in yeast tablets, liver, kidneys, wheat-germ) and protein (eggs, meat, fish, cheese) for a couple of weeks before that colouring session. Here's a guide to colour and shine treatments you can do at home :

### Fair Hair

If your hair is mousy and you want blonde highlights, use a model-girl trick : sit in the sun, stroking strands of your hair with half a lemon, squeezing well to make sure the juice covers the whole strand. The natural bleaching action of the sun and lemon juice will lighten it most effectively. Camomile flowers are excellent for lightening hair. Steep four tablespoons of dried flowers in 550ml (1pt) water for two hours. Strain and use as a rinse, swabbing the hair thoroughly with cotton wool soaked in the brew. Alternatively, you can make paste by mixing one cup of camomile brew to half a cup of kaolin. Apply to hair for about thirty minutes, then rinse off thoroughly.

### Brown Hair

Light-brown hair can be lightened slightly with one of the treatments above, or darkened with a strong infusion of sage leaves. Steep four tablespoons of sage leaves (dried) in 550ml (1pt) boiling water for two hours. Strain and apply thoroughly to the hair after washing.

You can now buy natural henna products which enhance all shades of

Lemon juice to lighten fair hair

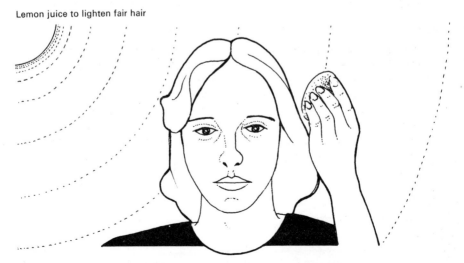

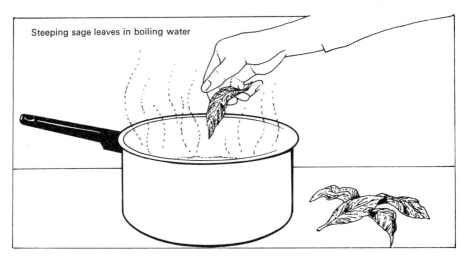
Steeping sage leaves in boiling water

brown hair, giving brown, chestnut, red or golden tints. Henna is one of the oldest known natural hair dyes. It is made from the crushed leaves of the lawsonia bush, a type of privet. Follow the instructions carefully, and you'll find that your hair becomes much more shiny, as well as a super colour, because the henna actually coats the hair shaft. But the colour *is* permanent!

To add shine to dark hair, after washing your hair, add a tablespoon of cider vinegar to the final rinsing water.

### Red Hair

A henna treatment will deepen and strengthen the colour of red hair. If your hair has a few red lights which you want to enhance, try rinsing it with this brew: steep a dozen marigold flowers and a teaspoon of saffron in 550ml (1pt) boiling water, for two hours. Strain and use after shampooing. Sit in the sun for a while if you can.

### Grey Hair

Darkening grey hair can be a tricky business and the result is often too hard. Experiment first with a lock of hair. Try a rinse using strong, cold tea after washing. Apply very thoroughly to the hair, and allow to dry. Alternatively, steep a mixture of sage, marjoram and raspberry leaves in boiling water (about one tablespoon of each) for two hours, strain and use as a rinse.

The shine on your hair is impaired if the natural overlapping 'tiles' which form the shaft are disturbed in any way. So always treat your hair gently if you want it to look shiny. Use a soft brush, a non-scratchy comb, and have the ends cut regularly. Wrap the brush in an old silk scarf and stroke it gently for shine.

For a dramatic, shiny look on the beach, which actually protects your hair and conditions it, comb olive or corn oil through your hair and wrap the tresses around your head. Swim and sunbathe all day, then shampoo the oil off at night.

# Get Fresh

Recently there has been concern that we are rapidly becoming a nation of deodorizing fanatics, using anti-perspirants with all the fervour of a hospital cleaning squad. Body-freshening products are becoming more and more expensive. Obviously, if you have a body-odour problem, you need an anti-perspirant, but you should only use it once a day and you should make sure that your diet and clothing aren't aggravating the problem. Most of us can be fresh with regular washing, dietary care and home-made deodorants. There is just no need to spend huge sums on anti-perspirants and subsequent cleaners' bills.

First, you must realize that humans need to perspire. We have around two million sweat glands—the eccrine glands, especially active on palms of hands and soles of feet, but present all over the body—which sweat fairly continuously, and the appocrine glands which are mainly confined to the underarm and pubic regions and sweat intermittently according to the emotional state of the person. The appocrine glands secrete a milky fluid which is attacked by bacteria, decomposes and causes BO. The eccrine glands secrete a clear liquid (water plus salts) which doesn't decompose but does support the bacterial growth started by the appocrine sweat.

So, for freshness, you do need an all-over wash once a day. If you can't take a bath, have a strip-wash, adding a tablespoon of cider vinegar to the final water. Home-made lavender water (see p. 41) is also a good deodorant. Add some to your bath-water or dilute it in a bowl of water and swab all over with a sponge.

Both these remedies are ideal if you are allergic to commercial deodorants. The vinegar is particularly good as it restores the normal acid mantle to the skin. When chrysanthemums are in season, use the leaves to make this deodorant wash : steep a cup of leaves in two cups of boiling water for half an hour. Mash and strain. Use the lotion to bathe underarm area.

You can also use chlorophyll-rich greens as a deodorant rub : chop very finely a mixture of parsley, watercress, spinach, lettuce, etc, or any dark greens. Tie in a muslin bag and rub it under your arms after bathing.

Many people who suffer from excessive BO problems don't realize that it's the food they are eating which is mainly responsible. Too much meat or fatty foods will cause bad breath and BO. This has been discovered by many slimmers following a high-protein, low carbohydrate diet. They get thin, but lose all their friends in the process ! So, cut down fatty meat, and spicy foods such as curry, and step up the natural body sweeteners like parsley, watercress and other salad foods and drink plenty of water. You should also drink a cup of sage tea every day : infuse a tablespoon of chopped sage leaves in a cup of boiling water for a few minutes, strain and drink.

Ancient man had no BO problems because his perspiration evaporated before bacteria had a chance to form. Always make certain that the clothes next to your body are in natural fibres—

cotton bras, pants, and shirts for preference. Synthetics trap sweat and produce the perfect environment for bacteria. After bathing, reduce the heat under your arms by rinsing that area with cold water and cider vinegar. Allow to dry naturally, then dress. If you feel you need to top up with a commercial product, use a roll-on lotion. These give greater coverage than aerosols and cause less waste.

However, you must let them dry thoroughly before dressing.

In the pubic area, hygiene is obviously vital. Wash thoroughly, and wear cotton panties for maximum protection. Nylon pants topped by tight jeans will cause odour. You should not need a vaginal deodorant with careful daily hygiene. Nowadays vaginal deodorants are considered not only unnecessary but also possibly harmful.

# Massage for Beauty

Massage is a beauty treament that rich women pay pounds for, but it can be yours, for free, at home. Massage softens your skin, helps you to relax and generally revitalizes you. It can't slim you down, but it can certainly pep you up. When you have a quiet evening, try this easy routine. It's a self-massage, using a do-it-yourself massage oil.

## Body Massage

### The Oil

Add four cloves and the peel of half a lemon (cut in shreds) to a cup of corn oil, and leave it overnight. Remove the cloves and peel before using.

All you will need is the oil, an old towel and a warm room. Tuck your hair into a shower cap, remove your clothes and spread the towel on the floor. Sit on the towel, and smooth a little of the oil into your thighs with both hands. Massage firmly towards your knees, then pinch the fat areas with finger and thumb.

Add a little more oil to the palms of your hands and massage your arms, from shoulder down to finger tips. Take especial care with your hands, working the oil between fingers and into your nails.

Kneel up and massage your tummy and midriff with firm strokes—right hand moving from the left side to the right, left hand moving from the right side to the left. Knead fatty area with your knuckles, then make 'S' folds using thumb and forefingers of both hands (see illustration) all over the area. Still kneeling, massage buttocks in firm circular movements, then knead them firmly with your knuckles.

Using more oil, smooth your body from the waist up over your bosom to your shoulders, using gentle upward movements. This is most important; you should never massage the bosom downwards as this could give your figure the droops!

Stand up, and, using both hands, rub each side of your spine, reaching over your shoulders as far as you can.

Place one foot on a chair or stool and massage your feet, rubbing the oil between your toes and using circular movements on the balls of your feet. Work up to your knees in smooth strokes.

By now, your body should feel relaxed and glowing and the oil should be almost absorbed by your skin. Blot off any surplus and take a long, relaxing nap lying on the towel on your bed, or slip into a warm scented bath.

## Face Massage

This is a more delicate process which

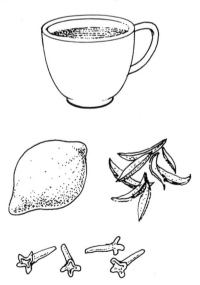

should be done with great care, using the lightest possible touch.

## Cream
If you have a dry skin use a favourite moisturizing or night cream or a light oil such as almond oil. Don't massage your face if your skin is very greasy as the action will stimulate the sebaceous glands, which are already working full time to produce sebum.

Take off your sweater, shirt or dress to leave your neckline uncovered and tie your hair back. Put on a hairband or curler cap to protect front and side hair.

Smooth the cream into throat and massage upwards to your chin with firm strokes. Pat a little cream on your nose, smooth upwards and outwards

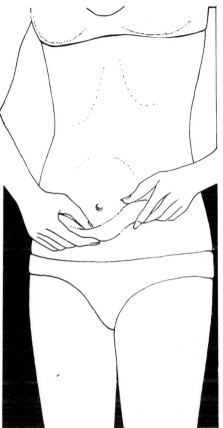

Massaging 'S'-shaped folds

Massaging feet

over your cheeks towards your hairline with the pads of the second finger of each hand. Use the same movement to smooth cream from chin to ear-lobe level. Massage the area between your brows, upwards towards your temples, in the same way. Repeat each movement ten to twenty times.

# Problem Areas

Do you suffer from rough skin on your knees, ankles and elbows or a dry neck ? Then give these areas a little extra attention when you take your daily bath. You'll find that they will improve quite quickly.

## Knees

First, make sure, if possible, that you don't add to your knee problems by standing on your feet all day. Puffiness can also be caused by fluid retention in the legs. Beat this with a healthy diet, daily resting (putting your legs higher than your head) and massage. Tackle rough skin with oil massage— use lemon-peel oil made by turning half a lemon skin inside out and filling it with olive oil or corn oil. Leave this overnight. Then warm it and rub it well into the dry skin.

You could also try this mixture : beat an egg-yolk with half a cup of milk and a tablespoon of oil. Massage the mixture into your knees, leave for fifteen minutes (while you wash your hair, perhaps), then rinse off.

## Neck

Dull, dry-looking necks are often caused by the excessive wearing of polo-necked sweaters in winter. The wool or synthetic yarn provides a snug cocoon, but it doesn't allow the skin to breathe properly. The result is a dirty, greyish-looking neck. Leave off your sweaters occasionally, and give your neck the following treatment in springtime.

### Massage

Every night rub warmed corn oil into your neck, from throat to chin level. Blot off any surplus before you go to bed. Every week make a slightly richer brew by mixing a cup of warmed oil with one egg-yolk and a tablespoon of brewer's yeast. Pat it into the skin and leave it for fifteen minutes. Then rinse it off with tepid water.

Bandaging ankle with leeks

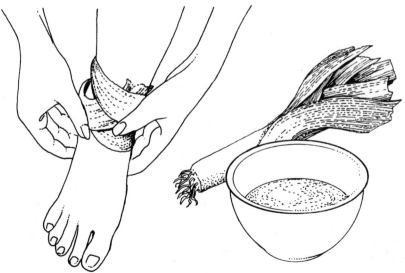

### Exercise

Try strengthening neck muscles by saying 'yes' fifty times before breakfast! Nod up and down fifty times and make fifty circular neck movements, first to the left, then to the right.

### Fresh Air

Tie your hair back, put on a low-necked sweater and take a long, healthy walk. If the weather is too nippy for such treatment, simply tie up your hair, wear a round-necked jumper and rub a little moisturizer into your neck. Avoid all rough material next to your throat for a few weeks.

## Ankles

Use the knee treatments above and one of the following if your ankles are rough, mottled-looking or itchy.

### Leeks

When you are cooking leeks for the family, save the water, or a softened leek skin, and use to bathe or bandage the area. Leeks are also good for corns. You can use onions in the same way, although the smell is somewhat stronger and some people may find it unpleasant.

### Porridge

Oatmeal is soothing and healing. Use it as a paste mixed with milk. You can add honey, too, if your ankles are very red and painful. Smooth on, leave for fifteen minutes, then rinse.

## Elbows

We often ignore elbows because we can't see them—the trouble is that other people can. If you wear a sleeveless evening dress or summer shirt, the effect can be ruined if your elbows look sore or knobbly.

Try resting your elbows in a squeezed half lemon for a few moments. Rub baby or olive oil into your elbows regularly after your bath, or use a little home-made mayonnaise on them before bathing. Other elbow-soothers include porridge, avocado skins, honey and egg, cider vinegar and water (one tablespoon to one cup of water).

# Beautiful Teeth, Fresh Breath

If your teeth are white and your breath is sweet, then your smile can be one of your greatest beauty assets. But, sadly, most of us really only start to care for our teeth and gums when mouth troubles become a big problem. Try thinking of careful tooth care as an insurance against a painful old age as well as part of your beauty routine. That way, you'll be sure to remember those thorough cleaning sessions.

## Diet and Your Teeth

Plaque, the invisible film of bacteria which leads to tooth decay, builds up around your teeth very quickly and must be removed at least once every twenty-four hours to give your teeth a fighting chance of staying healthy. Brush thoroughly between your teeth, too, and where your teeth meet your gums. Invest in a new toothbrush every couple of months, and use dental floss

(wax-coated twine, inexpensive and obtainable from most chemist's) to clear away food debris. In your diet cut sweets and sweet sugary soft drinks to a minimum. If you can't do without them have them with a meal and clean your teeth thoroughly afterwards. Natural exercise for teeth and gums comes from chewing. Roughage such as raw carrots, celery and apples is good for this. Try to include a salad of raw vegetables at one meal each day. Dentists are now more in favour of ending a meal with cheese than with an apple. The cheese leaves the mouth in a bland, non-acidic condition which is less likely to aid corrosion than the acid left in the mouth after eating an apple. So, if you can't clean your teeth, do nibble cheese instead.

## Natural Cleansers

A simple tip is to drink water after and between meals to help dislodge particles of food and leave your mouth feeling fresh.

Sage is the most useful herb for the mouth. Rub sage leaves on your teeth to clean them and help remove stains ; the sage oil contained in the leaves is naturally cleansing, yet gentle and non-acidic. The sage also leaves a lovely fresh taste in your mouth. Rub the leaves on gums too, to clean and strengthen them. Use 550ml (1pt) boiling water to 28g (1oz) dried sage or half a cup of leaves, steep for fifteen minutes, then strain and use the tea as a gargle and mouthwash. (Try this if you have any mouth troubles, from ulcers to bleeding gums.)

Lemon peel is cleansing, too, but because it's an acid you must rinse your mouth very thoroughly after

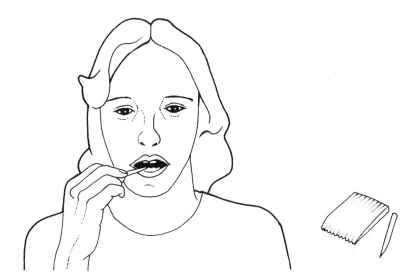

rubbing your teeth with the pith side
of the peel.

Crushed strawberries make a good
natural tooth cleanser. Rub half a
strawberry against your teeth, or dip
your toothbrush in squashed
strawberries, then clean your teeth.

Salt and water made into a paste will
remove stubborn stains caused by wine
or smoking, but use sparingly as this
can damage the enamel with over-use.

Honey and bread charcoal (the
scrapings of burnt toast) is another
winning combination for your smile.
Blend a teaspoon of honey with a
teaspoon of charcoal. Dip your brush
into it and brush your teeth in the
usual way, then rinse your mouth with
cold water.

## Mouthwashes

Sage tea is a splendid mouthwash if
you have mouth troubles. If bad breath
is a problem or you have been eating
spicy, garlic-laden food, add a drop of
peppermint essence to a glass of cold
water and rinse your mouth

thoroughly.

This recipe makes a strong
mouthwash which should be diluted
in water : mix 140ml ($\frac{1}{4}$pt) cooking
sherry, a teaspoon each of powdered
cloves and nutmeg and a pinch of
ground cinnamon. Shake in a bottle
and add two or three drops to a glass
of water when needed.

When lavender flowers are in bloom,
gather a cupful and add to 550ml (1pt)
boiling water. Steep for fifteen
minutes, strain and use the liquid when
cool as a mouthwash—it's fragrant and
refreshing.

## Instant Breath Sweeteners

Chlorophyll-rich parsley, watercress,
beet and turnip tops are good for your
breath and complexion. Chew parsley
and watercress or use them mixed in
salads. They combine well with other
vegetables for a beauty drink : in a
little water liquidize or stew vegetables
together until soft, then sieve, cool and
serve. Chewing a clove will instantly
refresh your mouth and breath.

# Beauty that Travels Well

In this high-speed age, most of us have to do some travelling—the daily commuter journey to and from work, or the annual holiday. Journeys themselves, and short or long stays in different places can test your beauty routine. It's hard to look fresh and attractive out of your own schedule and environment. But there are certain easy rules you can follow—and they don't involve spending a lot of money on special products and equipment.

## Getting There

On a short train journey (the daily commuter train to work, for instance), protect your skin from smuts and grime with a complete make-up before you set off. Tuck in your bag a little plastic box, full of squares of lint or muslin soaked with rose-water, and use them to freshen your face and cool your brow before you arrive and to remove the inevitable smuts on your nose.

If the journey is longer (by train, 'plane or boat), start off with a light make-up—foundation, eye make-up, lipstick—but no powder. In pressurized cabins on aircraft, make-up quickly gets clogged and goes shiny. It is best to retouch and powder just before landing.

An aerosol spray, containing rose-water or mineral water, is a good freshener (not be used on flights though ; use instead the lint squares). It can be sprayed over make-up if there's no chance of washing. Eau de cologne or lavender water are the most refreshing perfumes for your journey. More exotic smells tend to go stale with the smoky atmosphere.

On any long journey, keep your hair-do simple, save hair-washing and setting for the evening you arrive.

Perspiration is another problem—however icy the weather outside, your transport is probably going to be over-heated. Wear cotton undies and a fresh cotton T-shirt next to your skin. Tuck another into your bag to change into just before you arrive or mid-way through the journey. Trousers or jeans, if they are tight, are often very uncomfortable to travel in—choose an uncrushable skirt instead and wear comfy shoes. Air hostesses who have to be chic all the time, despite air pressure, overheated cabins and the rush of coping with meals, wear loose shoes and clothes to counteract the swelling that air travel produces.

If you have a very long trip, sleep is a good—and beautifying—way to pass the time. Ask for a milky drink or take a couple of calcium tablets to help you relax. Avoid alcohol which may make your kidneys over-active and give you a hot and bothered feeling.

### Sitting Down Exercises

Try these exercises in a car (not while driving), boat or 'plane to tone your muscles and help you 'stretch your legs' :

1 Sit with your back straight, tummy tucked in, feet together. Now push your elbows back hard ; try to make your shoulder-blades touch behind you. Hold briefly, then drop elbows and raise shoulders. Repeat five times.
2 Place your handbag between your knees and press them together. Hold briefly, then relax. Repeat ten times.

3 Stretch your legs out in front as far as possible, point your toes forwards, then towards the ceiling. Repeat twenty times.

## Arriving

Remember that travelling abroad isn't the only type of travel that upsets your beauty routine. Water, food, pollution, environment all change quite radically from one part of *this* country to another. So be prepared.

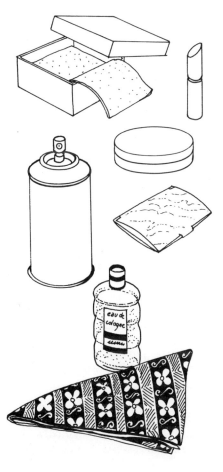

### Diet

If local food is more spicy than you're accustomed to, go easy on those exotic dishes. Your system could react by giving you a crop of spots or a nasty rash. Drink lots of mineral water, choose plain grilled meats and fish, fresh fruit and vegetables. If you're on holiday, avoid overloading your body with three huge cooked meals a day if you're used only to one. Eat smaller portions at each meal.

### Beauty Routine

If your skin is normally very dry and you're going to a colder climate then soothe your face with a moisturizing cream and try this facial : mix a tablespoon of olive oil with a tablespoon of yoghurt. Pat it on your face and neck and allow to dry for fifteen minutes. Remove with tepid water. If you're holidaying in a hot place, you must protect your face with a good anti-burn lotion. At night, use warm olive oil, patted on to your face and neck to soothe the skin and help counteract the sun's drying action.

Greasy skin benefits from the sun, but you should still use a moisturizing cream during the day. If you go to an exotic place, use the cheap, local fruits for facials : pineapple juice, melon pulp, mashed strawberries, orange peel (zest side), diluted lemon juice, mashed cucumber and apricot pulp are all good for greasy skin.

Take a little plastic bottle of cider vinegar with you to mix with local water for a final rinse of face and hair. This ensures a soft, soothing, acid-balanced result. If the water is very hard or brownish in colour, use mineral water for face washing.

# Slimming the Natural Way

You don't have to supplement your diet with slimming foods to lose weight —you can do it simply, and healthily, by understanding what effect various natural foods have on your body. You don't need expensive equipment to tone your muscles either—just willpower. What's more, doctors have recently started promoting another way of controlling the awful eating habits which make you fat, for no cost at all. It's called behaviour therapy. Here's a guide to all three slimming methods showing how they can work together to trim your weight problem.

## Diet

Many women suffer from fluid retention which can make them feel and look puffy and bloated—especially before their period. One of nature's own diuretic substances is found in citrus fruits, ie grapefruit, lemons and oranges. Eat half a grapefruit for breakfast, an orange at lunchtime, and sip lemon tea if you have a fluid problem. The rest of your diet should consist of fresh produce and lots of water to drink. Go easy on dairy foods, coffee and tea and resist the lure of sweets, starches, cakes and sugars.

If your problem is nibbling between meals just alter your eating pattern to include a hearty breakfast. It should consist of protein in the form of eggs, fish (kippers, for instance) or meat, plus a slice of wholemeal bread and black coffee or tea. The protein will give a long, slow boost to the blood sugar levels which control your hunger pangs and you won't want those nibbles. If possible, eat your other big meal of the day at lunchtime and cut the evening meal right back to a simple snack, like yoghurt or fruit. You shouldn't feel like anything bigger.

If you are a sluggish type without energy and feel that your body metabolism isn't burning off surplus foods fast enough, then speed things up a little. Try eating six or more small meals during the day, instead of one large meal. This will help you to slim more quickly. Each mini-meal should have protein (an egg, a small piece of cheese, etc), green vegetable or fruit, and something to get your teeth stuck into like a crispbread. But you must make sure that your daily calorie total adds up to no more than 1,500.

If your problem is lack of slimming discipline and a job that demands entertaining, then try a weekly mini-fast. Eat eight oranges and drink unlimited water for one weekend day when you are not being social. This will help give your body a rest. But you should make sure that those daily meals are low on starches and animal fats, and high on fresh vegetables, fruits and simple fish dishes. Alcohol is the worst fattener of all because it doesn't feel fattening until it's too late. Always ask for low-calorie tonics and mixers to help halve potentially high calorie totals.

## Exercise

Walking is yours for free, and it's great exercise. Jogging is good too, but don't embark on a daily jogging programme unless you're already physically fit. Instead, jog for twenty-five paces then walk for a hundred, gradually increasing the jogging paces over a few weeks. Try the following programme each morning and evening for spectacular results:

1 Lie on the floor, hands by your sides and raise your legs slowly off the floor to vertical position. Now lower them to a slow count of ten. Repeat ten times.
2 Turn over on your tummy. Raise your left leg as high as it will go, cross it over your right leg to touch the floor. Repeat with right leg, crossing over left leg. Repeat the whole movement ten times.
3 Stand up, feet apart, hands on your waist. Bend to the left, then to the right without twisting your body. Repeat twenty times.
4 Swing your arms backwards singly, then together, for a few moments.

This quick, easy routine helps to strengthen and firm tummy, calf, buttocks, waist, midriff and arm muscles and also gives your bosom a valuable lift. It takes just five minutes.

## Behaviour Control

The cleverest way of all to slim is to control your own eating pattern by keeping temptation out of your way and learning to enjoy food (lots of fat people don't—the food passes from mouth to stomach far too quickly).

Chew each mouthful twenty times and enjoy the taste before swallowing. Take your time. Eat the food on your plate slowly and deliberately, finishing the meat or fish first, the vegetables next and keeping the different types of foods separate in your mouth.

Always eat sitting down at a place laid at the table, and use small plates. Make a mental note that you are not allowed to grab odd biscuits out of a tin or eat casually in the kitchen. Prepare small portions for you and the family in individual dishes, not a large amount in a communal pot as this will invite second helpings.

Keep biscuits and cakes in tins high up in the larder, never on display where you can see them. Be busy. Knit, sew, decorate the house, learn to drive, dig the garden, take on lots of time-absorbing projects while you slim. Avoid confectionery and baker's shops. If you must shop at a supermarket, do so after a good meal and make sure you're armed with a list.

Take your lunch to work so you're not tempted to join the hamburger brigade. Treat yourself to an ego-feeding free treat like a walk in the park or an hour's window-shopping every day.

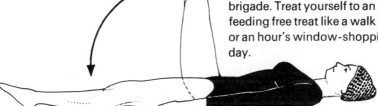

# Tanning Nature's Way

If you want a smooth, even tan, you should start thinking about preparing your skin before your holiday, especially if you have sensitive or fair skin.

A whole month before your holiday, start increasing your daily intake of the foods that will help your skin cope with the sun : Vitamin A foods (carrots and tomatoes), calcium (milk and cheese), vegetable oils, and Vitamin B foods (liver and yoghurt). Try eating a crisp, tossed salad every day with a vegetable-oil dressing, grated carrots and a natural yoghurt to follow. This strategy will help your tan.

At the same time, increase the lubrication you give your body by adding extra baby or corn oil to your bath-water. Rub plenty of moisturizing lotion into vulnerable areas : behind calves, feet, upper arms, neck and cleavage.

If possible, catch a little sunshine before you go, using sun oil containing natural bergamot, a fruit which helps speed up the sun's effect and lets you tan more quickly.

In the sun, make sure you eat plenty of oranges as Vitamin C will help prevent your tan becoming blotchy, and keep up your daily intake of Vitamin A and oily foods.

If your skin is sensitive, use a good commercial sun-screen. After sunbathing, rub your lips with honey, and pat your face with yoghurt, then rinse off. Any sore spots can be soothed with mashed cucumber, grated potato, strong tea or diluted vinegar. Make sure you tan gradually, working up from five to ten minutes daily to about thirty minutes' sunbathing. Be especially careful to reapply sun lotions after bathing, and avoid the really hot midday period. Take a siesta or a walk away from the beach or

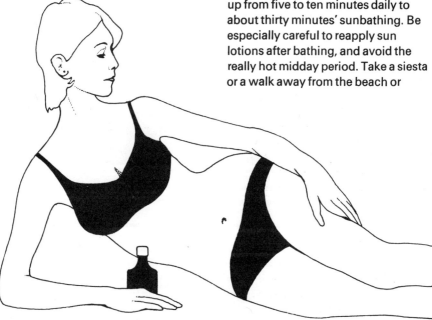

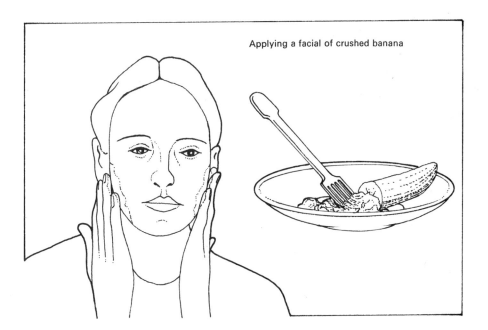

Applying a facial of crushed banana

linger over a long, relaxing lunch. Don't aim for a deep brown tan. It won't suit you, and will take painful time to acquire.

If you tan easily, use a simple brew of olive oil and vinegar as a tanning lotion (mix in equal parts), but still take things slowly at first. Be especially careful if you are on the Pill or take any kind of medical drug, as these could make you more sensitive to the sun. Pill-users, however, normally find that their skin tans slightly more easily.

After sunbathing, shower with lukewarm water to remove oil. Pat your skin dry, then apply a little light oil such as baby oil. But be sparing, otherwise you could make bubbles of trapped oil beneath the skin.

To prolong your tan, at home, keep up the skin-soothing oil massage treatment and add extra liver, kidneys and oranges to your diet. Your tan

actually flakes off with the tiny particles of skin that are shed every day. Your face has more exposure to the air so tends to become pale before your body. Avoid wearing make-up for as long as possible, just use moisturizer, and get as much sun and fresh air on your face as you can. Feed your complexion with a daily facial of crushed banana or grated carrot and oil, yoghurt, or egg-yolk and oil.

# Give Yourself the Waterworks

Turn on your tap for an inexpensive, yet effective beauty treatment. Water is a highly underestimated aid to good looks. Here are some bright ideas for using the substance which makes up about 70 per cent of your body anyway.

### In Your Diet

Drink about eight glasses of cold water a day to get your kidneys working efficiently and help keep your skin clear and your eyes shiny. If you have an elimination problem, drink a long, cool glass of water on waking to set things in motion nicely. In hot weather, mild dehydration can make you feel low and grouchy so you must replace the water lost in perspiration. You should drink at least 550ml (1pt) fluid for every 5.6°C (10°F). Try replacing those endless cups of tea and coffee with mineral water—you'll feel and look much better.

### For Your Face

Washing with soap and water is an old fashioned but extremely effective beauty concept and one highly recommended by dermatologists unless you have very dry skin. Do it like this: Splash your face all over with warm water. Lather soap (choose a bland, unperfumed brand) on your hands and rub gently on your skin concentrating on the oily 'T' area over forehead, nose and chin. Rinse thoroughly.

To get the circulation going and wash away tiredness, first thing in the morning, splash you face repeatedly with cold water. Puffiness and open pores can be reduced by rubbing an ice cube over your skin.

For tired eyes, soak a muslin pad in iced water in the fridge for an hour or two, then wring it out and place it over your eyes while you rest for fifteen minutes. This will rapidly restore sparkle.

Set make-up or freshen it with a fine spray of mineral water.

### For Your Body

Try a refreshing shower to stimulate circulation and make you tingle all over. First, have the water pleasantly warm, then gradually have it cooler until it's really cold. Afterwards, wrap yourself in a fluffy towel and rub briskly for a few minutes. You can get the same effect in a bath by starting off with warm water, and letting the cold tap run while you soak.

To help firm your breasts, use alternate showers of warm and cold water, spraying upwards with a hand-spray attached to the bath or basin taps. Use the same hot and cold treatment to put new life into weary feet.

### For Your Hair

Rinsing your hair in rainwater makes sense if you live in the country (well away from industrial pollution) and if the water is softer than the tap variety. Test it like this: collect a jar of rainwater, add a teaspoon of shampoo to it and shake hard. Do the same thing with a jar of tapwater. If there is less lather in the tap water, your water is hard. Collect rainwater and use it for that final rinse to leave your hair feeling soft and silky.

## Underwater Exercises

1 Lie back in your bath, resting on your forearms and elbows at the sides of your body. Slide your feet back so that they are flat on the bottom of the bath. Raise your hips and swing them gently from side to side. Keep it up for a count of twenty. This does great things for your hips and waistline.

2 Lie back, feet straight out in front, hands palm down on the bottom of the bath. Now draw your right heel back as far as possible towards your bottom. Straighten your right leg and start drawing up the left heel at the same time. Keep it up for a count of ten. This is good for tummy, bottom and thighs.

3 Slide down in the bath until the water covers your bosom, anchoring your feet firmly against the tap end. Put your hands out in front of you, palms down, with your arms floating on the top of the water. Turn palms outwards and press firmly outwards and downwards. Repeat ten times. This is good for your bust and midriff.

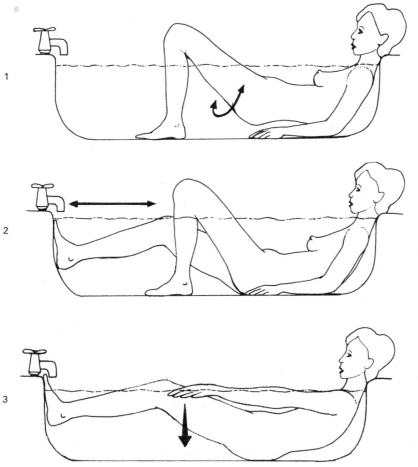

# Free as Air

Fresh air is a natural cosmetic which is yours for free. However, most of us don't get the full benefit of this beauty aid because we spend too much of our time cooped up in stuffy rooms or carbon monoxide-poisoned streets and take such shallow breaths that our bodies are starved of fresh air. To stay beautiful, make time for some deep breathing and improve your breathing action. If you can, enjoy fresh air at the seaside or in the country at weekends.

## How to Breathe Correctly

First, make sure bad posture isn't cramping your breathing style. Diaphragm, rib-cage and lungs need plenty of room for action. Stand or sit up straight with shoulders relaxed, backbone in its natural 's' curve— your body should be stretched as if to look taller. Never hunch, particularly when walking up-hill. You'll be puffed and breathless in no time. Now try these breathing exercises:

1 Stand with legs apart. Place the flats of your hands over your tummy. Now breathe in and out very quietly and gently making sure that you can feel your stomach moving in and out. Increase the rhythm of your breathing action, taking longer breaths and feeling your tummy move more deeply. Go back to the gentle action for a while, then relax. Repeat this exercise twice a day to improve your breathing technique.

2 Sit on a chair at a table, arms folded over a cushion or pillow. Lean forward, but keep your back straight. Breathe deeply and evenly until you

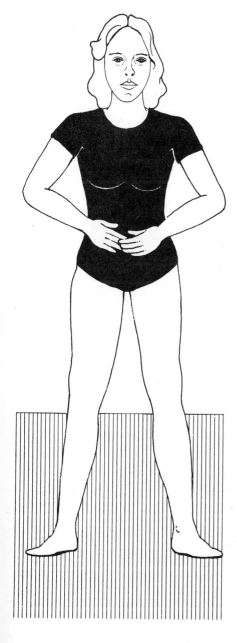

Breathing exercise 1

feel really relaxed. This is a wonderful exercise at the end of a busy day. It will help you feel and look sparkling fit again.

3 Sit comfortably, back straight, eyes closed. Close your right nostril with your right forefinger and breathe in deeply through your left nostril. Now release your right nostril and close your left nostril with your left forefinger. Breathe out through your right nostril. Repeat. This is a soothing, decongesting exercise, good before an important meeting or an exam. The oxygen gives a boost to the brain as well as the body.

It's good for your complexion to breathe free of make-up occasionally, especially if you have a greasy skin. The skin needs oxygen to promote the cell-renewal process, which is taking place in the epidermis, and to speed up the removal of the debris of old skin cells which can make your face look muddy. So, to help your complexion, go for country walks wearing just a little moisturizer on your face.

Use the air in your garden to dry your hair naturally after a shampoo. More and more hairdressers are advocating natural-drying styles and recognizing that too much heat is bad for the hair, causing brittleness and split ends. Wash your hair before breakfast in summer and sit outside while you eat, letting the air work for you.

On a blustery day, wrap up well and protect your face with moisturizing cream. Walk briskly into the wind to make your muscles work extra hard. You'll feel tingly and exhilarated afterwards.

# Down-to-Earth Beauty

Two of the top ingredients for natural beauty recipes are fuller's earth and kaolin. Both can be purchased from any large chemist's.

## Fuller's Earth

This is a kind of clay which possesses highly absorbent properties. It comes from various deposits in America and from Redhill, Surrey, England. Originally, the substance was used for 'fulling' (cleaning) cloth, hence the name. It contains minerals which help to improve the skin's circulation but if applied neat, it can be very drying indeed so it is normally used as part of a mixture of ingredients. The mineral which gives fuller's earth its absorbent action is called hydrated aluminium silicate (also used in formulating modern anti-perspirants). Grease in the skin coats the particles of this substance and is lifted away when the fuller's earth is removed. So fuller's

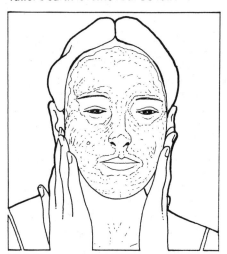

earth is really more suitable for greasy skins or as the fixing powder for a blend of fruit or vegetable ingredients in a mask for normal or dry skins. Never put fuller's earth neat on to a dry skin. But do experiment with it in your own natural cosmetic repertoire.

Try these recipes : Mix half a teaspoon each of Friar's Balsam, glycerine, and rose-water with 56g (2oz) fuller's earth. Apply to your face, allow to harden, then rinse off with tepid water. This should boost circulation and thoroughly clean the skin. Recommended for greasy, sluggish or muddy-looking complexions.

If you have a foot perspiration problem, try mixing your favourite talc with an equal quantity of fuller's earth. Sprinkle on to your feet and rub in after your bath and before dressing. You'll find that your feet will be odour- and perspiration-free for hours.

## Kaolin

This is a fine, grey clay used to make delicate pottery and it can help to give you a porcelain complexion. It's formed by the decomposition of the feldspar minerals which make up a large proportion of the rocks of the earth's crust. However, kaolin is only found in a few places—China, Japan, Devon and Cornwall, Limoges in France and some parts of the United States. Like fuller's earth, it possesses drawing qualities, removing grime and dead cells from the skin. But it is very astringent, so don't use it neat unless your skin is very greasy.

Try using kaolin as a binding substance for facial masks. For dry skins mix it with any of these

ingredients: milk, honey, egg-yolk, mashed banana, peach juice, almond or corn oil, cream, yoghurt. For greasy skins mix it with lemon juice, vinegar, apple purée, pineapple juice, cucumber (mashed or juice) or egg-white.

You can use them singly or together—depending on what you have in your larder. Remove kaolin-based masks with lukewarm water and let your skin settle before reapplying make-up.

## Mud and Peat

These have been associated with beauty since ancient times. Some health-food shops and big stores stock Moroccan mud which can be bought quite cheaply. It makes a superb hair conditioning treatment. Peat for baths, such as the mineral-rich substance from Austria, is also available from health-food shops. If you live in the country in a clay area, you can make your own mud-packs from clay dug up from the garden. But you must remove all stones first and boil and strain the mud several times to remove any impurities. Add a few tablespoons of fuller's earth to bind the substance before patting it on to your body. Allow to dry, then rinse off. Don't use garden mud on your face.

Other earth-bound cosmetic ingredients include daisies and dandelions. If your beauty is marred by bruises, boil up a couple of handfuls of daisies, heads and stems, etc, in 550ml (1pt) water. Let it bubble for ten minutes, strain and cool. Apply a cloth, wrung out in the mixture, to the bruised area for a few minutes to reduce the mark. You can drink the same concoction, sweetened with honey, to relieve a cough or catarrh.

Dandelions are useful as an internal cosmetic. The leaves are good in salads and they have a diuretic action which helps fluid-retention sufferers.

# Grooming for Men

Men are becoming increasingly aware of the need for good grooming—in fact, many are as firmly caught up in a daily beauty routine as women. But men can be vulnerable to the type of television advertisement that projects the virile image and they are sometimes influenced into using the wrong products for their skin and hair just because a famous personality uses them. Here are some ideas on how a man can look good the natural way.

## His Skin

This can be the greasy, dry or combination type and it won't be very much tougher in texture than a woman's. Indeed, shaving effectively removes the protective layer of sebum and skin cells every single morning, so his chin and neck areas are likely to be sensitive. A bracing after-shave will make things worse. Instead, persuade him to use a gentle, natural product like witch-hazel and rose-water mixed in equal parts. A moisturizer like baby lotion, or rose-water and glycerine would also be a good idea. He should make sure that his diet is rich in skin foods. Many men miss out on valuable Vitamin C because they prefer a pint and a sandwich at lunchtime, a cooked meal and pudding at night. Fresh orange juice or a peeled orange or grapefruit at breakfast-time will ensure a good supply.

## His Hair

Any of the recipes in the hair sections (see pp 26—33) will be suitable for a man, depending on his hair-type, of course. He should avoid harsh anti-dandruff shampoos which could make his hair very fragile indeed with over-use. For falling or thinning hair, a daily massage can help improve the circulation together with a vitamin-rich diet and plenty of rest. Stress is a big factor in the hormonal activity which triggers off alopecia (baldness). Trichologists agree that it is extremely difficult to reverse the trend once bald patches appear. However, a daily scalp massage is certainly good for all kinds of hair problems : simply place fingertips on the scalp and make circular movements, moving the scalp itself, not your fingers. Repeat all over the head. Many men drag a sharp comb

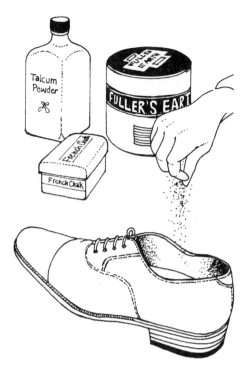

BO should be kept under control with twice-daily showering, eating a balanced diet with plenty of green vegetables, drinking water and washing underarms each day with a mixture of cider vinegar and water (two parts vinegar to one part water). Cotton vests and shirts are more absorbent and will help prevent the odour from becoming unpleasant. Bad breath, caused by eating rich foods and drinking spirits or beer can be controlled with a mouthwash. It also helps to chew parsley or watercress. Foot perspiration is another male problem. It should be treated with a daily water and vinegar rinse, cotton or wool socks only (never nylon), and a mixture of talc or equal parts of French chalk and fuller's earth should be sprinkled into shoes to keep them fresh-smelling.

The correct diet will keep his body trim and slim. It should also be remembered that, with the passing years, calorie intake should be decreased to correspond with decreasing exercise. Sporty men who give up sport—but not the eating and drinking that go with it—are in danger from heart trouble, respiratory problems and high blood pressure as they get older. It's relatively simple to adjust to a reasonable diet : cut out all stodgy and high-fat foods, keep drinking to a sensible level, and balance business lunches with very light suppers.

Many men take up jogging to keep fit, but this can be dangerous. Walking is better, gradually building up the jogging as you build up fitness.

through thinning hair several times a day, but they shouldn't. A comb should be carefully chosen to have well-spaced, rounded teeth. A soft brush is best for home use ; a baby brush is ideal if the hair is very thin.

## His Body

Men suffer from spots, rough skin and body odour just as women do, and any of the appropriate natural remedies in this book can be used. However, men often neglect such problems as spots or pimples on the bottom (often made worse by wearing nylon underpants), severe perspiration trouble, bad breath and rough skin. The spots will improve with a daily washing with bland soap and bathing with equal parts of lemon juice and rose-water. Cotton underpants should be worn.

# Day-time Beauty

It's hard to stay looking good all day if you work in an office, shop or factory where dust, dirt and extremes of temperature can combine to ruin your hair-do and make-up. If you are busy, then your appearance is the last thing you want to have to worry about —yet your appearance invariably matters. Here's a plan for good looks which stands up to the daily wear and tear.

Before you go to work, spend time making sure you look your best. Get up half an hour earlier, if necessary, to do your hair and make-up. Make-up will actually protect your skin from the dirt and grime of your journey. If it's windy or raining, tuck all your hair into a woolly hat (in pin-curls if you like) or under a smart scarf so that it doesn't look a mess when you arrive.

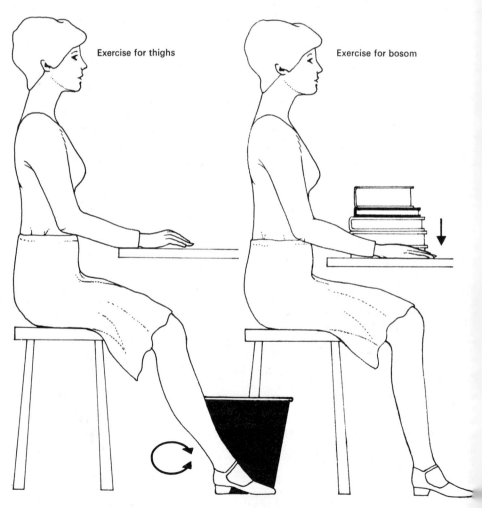

Exercise for thighs

Exercise for bosom

At work, try to keep a few personal toiletry items in your desk or locker. A natural bristle brush (saves the problem of carrying it to and fro in your bag every day), some bland soap (office soap is usually harsh), a bottle of rose-water with cotton wool for quick make-up freshening (cheaper than buying impregnated pads), a nail-file and bottle of nail varnish for emergency retouching, talc, eau de cologne, a tooth paste and toothbrush.

On arrival, brush your hair, reapply lipstick and freshen up with eau de cologne. Lipstick lasts longer if you top it with a little petroleum jelly and outline your lips with a pencil before applying it.

At lunchtime you should not need to do any radical touching-up, but tackle any smudges or dirty blobs with rose-water on cotton wool. Don't make the mistake of applying layer upon layer of pressed powder—your make-up will quickly become yellowish and stale-looking. If your nose is shiny, wipe gently with the rose-water, then powder very lightly. Do wash your hands and, if possible, clean your teeth after lunch. You'll be surprised how much fresher this will make you feel.

After work, if you are going on to a date, there is no need to strip off all your make-up. Simply deepen eye-shadow colours, apply more mascara, and perhaps a deeper shade of lipstick. Sprinkle talc in your shoes to make your feet feel fresh.

## Easy Isometrics

To avoid a spreading bottom if you sit at an office desk all day, try doing these exercises while you work.

They're called isometrics and they work on the principle of muscle contraction against an immovable object.

### For Your Bosom

Sit facing your typewriter or a large pile of books on your desk. Now place one hand each side of the typewriter or book pile, press hard and hold the contraction for a count of six. Rest and repeat.

### For Inside Thighs

Sit on a chair, arms relaxed. Now grip a wastepaper basket or similar object between the inner edges of your feet. Squeeze your feet together hard, count six, then relax.

### For Your Bottom

Sit on a hard chair (remove that soft cushion). Place your feet on the ground, and contract your buttock muscles hard. Hold for a count of six, then relax.

# Pick-Me-Ups

Sometimes, however disciplined your normal beauty routine may be, nothing looks right. For no reason, you feel jaded, run-down and depressed. Unfortunately, the beauty doldrums often strike just when you need to look good, before a party, for instance. Here are some ideas to boost your ego, revive your flagging spirits and put new life in you. Use them generously— they cost very little.

## Health Drinks

Honey is a good source of instant wellbeing—it supplies sugar for a lift and vitamins and minerals to make you feel good. Stir a teaspoonful into a cup of warm milk, add a squeeze of lemon or orange juice, stir and drink. You'll feel comforted and more relaxed.

Yoghurt is an excellent beauty food and drink. Whisk half a cup of natural yoghurt with half a cup of blackcurrant or orange juice. Sip slowly.

Beef extract is also warm, comforting and good for you. Make a cup of hot beef extract, and add a dash of sherry for an instant lift when you feel that everyone is taking you for granted. It's soothing and very calming.

## Resting

Lie down for just fifteen minutes with your feet on a cushion and a slice of cucumber over each eye, or a used, cold teabag if cucumbers are unavailable. You'll feel refreshed and your eyes will sparkle afterwards. Alternatively, lie with your bottom up against a convenient wall, your feet perpendicular to your body, for just five minutes. The blood will flow towards your brain, giving your feet a rest. At work, get the same effect by bending over in your chair and letting your head hang between your knees for a moment or two.

## Bathing

A warm, not hot, bath is refreshing after a busy day. Increase the stimulating value of your tub and give

Relaxed position in yoga

your tired body a boost by adding epsom salts or an infusion of mint leaves (steep a handful in 225ml ($\frac{1}{2}$pt) boiling water for fifteen minutes, strain and add the liquid to your bath-water). If your skin looks dingy, add an infusion of crushed blackberry leaves to your bath-water every night for a week to improve tone and texture.

## Walking

Put on some comfortable shoes and walk around the block, without the nagging extra of handbag or shopping-bag. Swing your arms for a change and take bold strides. You'll feel great when you return home.

For a simple energy-boost, take off your shoes and tights and walk around the house barefoot for half an hour. If it's warm enough, take all your clothes off and do it naked. The feeling of freedom is very refreshing, particularly if you've spent the day in hot, restrictive clothes.

## Yoga

Addicts say that there's nothing like yoga for a refreshing lift. Try this pose for just a few moments midday, mid-afternoon or just before you go out : lie on your back on the floor, feet slightly apart, hands a little way away from your sides, palms upwards. Now, let your toes flop outwards, close your eyes and breathe evenly. Try to imagine that you're getting heavier, heavier, heavier. Every bit of you is almost trying to push through the floor. Now relax totally for five minutes.

# Your Beauty Sleep

Nowadays, many people find it increasingly difficult to get a good night's sleep. Stress, the pace of life and many worries all add up to prevent those important hours of perfect, beautifying rest. Don't be a slave to sleeping pills if you have trouble switching off, instead try these ideas.

## Nightcaps

Calcium and magnesium are natural tranquillizers, so a milky drink really will help you sleep. Add a spoon of honey to sweeten the brew. If you don't like milk, then take a couple of calcium tablets with water or sip camomile tea. Chewing lettuce leaves (a natural soporific drug) can help, so can sipping mint tea or peppermint cordial.

## Relaxing Baths

A warm bath is sleep-inducing. Make it an aromatic one, by adding herbs (tied in a muslin bag to prevent any mess) or a drop of floral essence (rose is particularly pungent and soporific). Relax, then wrap yourself in a warmed nightgown or pyjamas and climb into a warm bed with one of the drinks described above.

## Yoga

If those daily problems just go round and round in your mind, try a totally relaxing yoga exercise just before bedtime. Lie on the floor, feet slightly apart with toes drooping outwards, hands and palms uppermost. Relax and breathe deeply. Try to imagine that your whole body is pushing down into the floor beneath you, getting heavier and heavier. Gradually, a sense of calm will come over you.

## Meditation

A simplified form of meditation can be

a great mind-relaxer. Light a candle and place it on a table. Now sit facing the candle and concentrate on the flame, emptying your mind of everything except the colour, warmth, shape and smell of that flame. Keep it up for as long as possible.

## Aromatic Relaxers

Smell can be a great sleep inducer, so try a scented bath as suggested above or dab a little perfume on your pillow. Make your own herb pillow by crushing a mixture of aromatic herbs in a mortar with some crushed, dried rose petals, some lavender or verbena. Add a few drops of Friar's Balsam, and sew the mixture into a little cotton purse made from an old hanky. Tuck it into your pillow case.

## Music

Never try to sleep after a late-night pop music programme. Instead, switch for just a few moments to some gentle, rhythmic, romantic music on another programme or on your record-player. If you have a portable cassette recorder, keep it by your bed with a soothing cassette ready for late-night use.

## Daydreaming

Just before you switch out the light, let your eyes wander freely around the room, focusing on each object and allowing thoughts to pop into your mind related to that object—memories, people, places. This will give your brain some fresh material to work on during the night, rather than those worries which are causing sleeplessness. Sex is also very good for relaxing your tensions.

Herb pillow

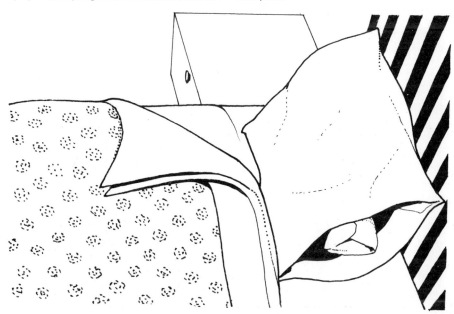

# Grow Your Own Beauty Parlour

To cut even further the cost of looking good, go back to nature by growing many of the ingredients used in this book. It is very simple to grow herbs in a window-box or pots if you don't have a garden. Even quite ambitious vegetables like cucumbers, used in so many natural beauty recipes, and tomatoes can be grown indoors. Here's a guide to growing for beauty. Don't be put off by experienced gardeners who say you can't do it—there's a conspiracy around to make gardening seem far more difficult than it is!

## Basil

Sow the seeds in trays in March. When the shoots are about 3.8–7.6cm (1½–3in) above the surface, dig up the seedlings very carefully (use an old spoon) and re-pot in plastic pots or sow in a warm, sunny, sheltered spot in the garden. The plants will grow from about 30–90cm (1–3ft) tall. To make them bushy, pinch off the shoots at the top as they appear. Basil will die back when bad weather starts around September, so use generously or dry it.

## Cucumbers

Select a climbing variety for indoor growing. Plant seeds or small plants in tubs around the middle of May, using plenty of water and feeding weekly with liquid manure. If you can, pick a spot by a window or glass door and arrange strings for the plant to climb up. They look most attractive, like an indoor vine. The flowers appear first; these need brushing gently with a paint-brush to transpolinate them. If you grow them outdoors, the bees will do this job for you. Keep up the manure feeds as the cucumbers start to form.

## Marjoram

Sow in trays during April and May, then transplant the seedlings to pots (indoors) or to a sunny outdoor border (this herb is very pretty in a country-style flower-bed). The plant grows 30–60cm (1–2ft) high and is a

perennial, so, if you choose a fairly tough variety such as common or pot marjoram, your beauty herb should reappear each year.

## Mint

If you can, plant mint in indoor pots and outdoors—it's such a useful herb that it's as well to have some at hand all the time. To encourage mint to take, plant a root horizontally, just below the surface in a well-dug, well manured spot, or indoors in a fairly big pot. Once the root feels happy, the plant will spread rapidly.

## Parsley

This useful plant can be sown from seed in boxes indoors or you can buy a root or two from your local nursery to plant later. Seeds should be sown from February, then transferred to pots or to a sheltered, well drained spot outdoors. Keep parsley well watered in hot weather. It's a biennial plant so will reseed itself in the second year but will have to be replaced in the third.

## Rosemary

An evergreen shrub which is pretty to look at in a border. It loves a dryish, sunny position, but is fairly happy even in poor soil. Buy in plant form and sow from September to March. Dig a hole a bit larger than the pot, place the plant in the hole and fill in by hand. Once the shrub is established, you can use the leaves all the year around, and dry them for indoor winter use.

## Sage

This attractive plant has a delightful smell. You can grow it indoors from seed (March planting) and then transfer it to a pot or border. Alternatively, you can buy plants and plant them direct as described above, in April or May. Pinch the mauvish flowers off to encourage more leaves to grow.

## Thyme

This is another aromatic herb which can be bought in seed or plant form. Sow the plants in a well drained, sunny position. Thyme likes a limy soil. You can also sow the plants in pots for your window-sill, but restrict the sun until the little plants are well established, otherwise they will shrivel. If your plant starts to get straggly, cut the leaves for your beauty and cooking requirements from the top of the plant.

## Tomatoes

These are a useful beauty food and can be grown indoors in a large tub. Choose the bush variety, as a single straggly stem on a stake looks unattractive and produces less fruit. Plant in May, in soil enriched with manure or compost, in a large tub which has a drainage hole at the bottom. Put the tub on a shallow container to avoid flooding the house whenever you water the plant. Choose a sunny spot, keep it well watered and feed with a tomato fertilizer. The plant will look very pretty in the hall or kitchen and the fruit will be useful. The bush variety grows to about 90cm (3ft).

## Drying Herbs

Cut longish branches of various herbs, tie together and hang upside down in a dry position—under the porch or in a cool airing cupboard is ideal—until thoroughly dried. Store in air-tight jars for winter use.

# Further reading from David & Charles

## GOOD FOOD GROWING GUIDE
**Gardening and Living Nature's Way**
John Bond and the Staff of 'Mother Earth'
A new-look growing guide to healthier and happier living
241 × 148mm   illustrated

## ECONOMY COOK BOOK
Mary Griffiths
A guide to how to cope with rising food and housekeeping prices and still produce tasty and nutritious meals
216 × 138mm

## COST-EFFECTIVE SELF-SUFFICIENCY
**Or The Middle-Class Peasant**
Eve and Terence McLaughlin
A practical guide to self-sufficiency, proving that life as 'middle-class peasants' is not only viable but enormously enjoyable and satisfying
247 × 171mm   illustrated

## EAT CHEAPLY AND WELL
Brenda Sanctuary
Rising food prices make this up-to-the-minute book a must for today's housewives
216 × 138mm   illustrated

## GROWPLAN VEGETABLE BOOK
**A Month-by-Month Guide**
Peter Peskett and Geoff Amos
A practical, easy-reference guide to growing super vegetables, and fruit too, month by month
250 × 200mm   illustrated

## GROWING FOOD UNDER GLASS:
**1001 Questions Answered**
Adrienne and Peter Oldale
An indispensable guide to setting up and maintaining every kind of glasshouse, together with an A–Z rundown of the familiar and unusual fruit and vegetables to be grown
210 × 148mm   illustrated

## GROWING FRUIT:
**1001 Questions Answered**
Adrienne and Peter Oldale
Answers all the questions a novice might ask about pests and diseases, choice of tree shapes and varieties, and pruning techniques
210 × 148mm   illustrated

## GROWING VEGETABLES:
**1001 Questioned Answered**
Adrienne and Peter Oldale
All you need to know about growing vegetables in a simple question and answer format
210 × 148mm   illustrated

## COMPLETE BOOK OF HERBS AND SPICES
Claire Loewenfeld and Philippa Back
A comprehensive guide to every aspect of herbs and spices—their history and traditions. cultivation, uses in the kitchen, and health and cosmetics
242 × 184mm   illustrated

## COOK OUT
Frances Kitchin
For the cook on a caravanning or camping holiday, Frances Kitchin provides the answers to all the problems when cooking meals with the minimum of facilities
210 × 132mm   illustrated

**British Library Cataloguing in Publication Data**

Voak, Sally Ann
    Natural and herbal beauty care.—(Penny pinchers).
    1. Cosmetics
    I. Title   II. Series
    668'.55      RA778

ISBN 0–7153–7550–4

Set in Univers
and printed in Great Britain
by Redwood Burn Limited
for David & Charles (Publishers) Limited
Brunel House   Newton Abbot   Devon

Published in the United States of America
by David & Charles Inc
North Pomfret   Vermont 05053   USA

Published in Canada
by Douglas David & Charles Limited
1875 Welch Street   North Vancouver   BC